# INNOVATION
## BY INDIA FOR INDIA

V. Ramaswami

ŚikshĀ
PUBLICATIONS, LLC
1049 Hillcrest Drive, Branchburg, NJ 08853, USA
email: sikshapublications@gmail.com

Ramaswami, Vaidyanathan
Innovation by India for India, The Need and the Challenge / Vaidyanathan Ramaswami

Includes bibliographical references and index.

ISBN: 978-0-9975777-2-3

ISBN: 978-0-9975777-2-3

**International Paperback Edition**
First published in 2016

ŚikshĀ

Siksha Publications LLC, 1049 Hillcrest Drive, Branchburg, NJ 08853, USA.
Webpage: http://www.innovationbyindiaforindia.com
email: sikshapublications@gmail.com

Printed at Gnanodaya Press, Chennai, India.

Cover design by S. Ravi Shankar
The following are acknowledged for the use of their work given out under the Creative Commons
Andrew Magill, Arun Katiyar, United Soybean Board, Sandra Henry-Stocker, Creativity103,
iamme ubeyou, Daniel X. O'Neil

# Preface

The number of Indians and persons of Indian origin who have enriched many nations, and humanity as a whole, through applied research and innovation in a multitude of fields has increased dramatically. This should be re-assuring for India particularly because a significant fraction of these have had their initial graduate education and training in India. There can be no doubt that the capacity of the Indian mind and the will of the Indian can match any other. India's indigenous scientific advances in the nuclear, space, and super-computer technologies are remarkable and attest further to the scientific and engineering talent in the country. Indeed, India's scientific and technical establishments have many feats to boast about, including the recent successful launch of an orbiter to Mars at a per kilometer cost less than that of a one kilometer auto-rickshaw ride in Ahmedabad.

Despite all of its above accomplishments, India is yet to harness commercially its research and innovation capabilities for its own benefit. Despite being a significant contributor to the information sector, not a single Indian enterprise has come close to any of the new age technology giants in terms of capitalization and profits. What are those impediments that hold back the Indian in India in the sphere of applied research and its commercialization? Why is it that new product generation is low even at the low end where little technology is needed? Can those issues be redressed and if so how? These concerns form the main focus of this book. Some key factors are identified and examined in detail in the Indian context to find some possible actions for improvement.

The literature on developmental issues as they relate to India is already extensive. Not enough focus, however, has been given to applied research and its commercialization, and it is in this respect that this book adds something new. My background has spanned academia, large industry, and a few start-up companies as researcher, manager, and strategist. I draw on this experience and the knowledge gained from my constant interactions with India to share some perspectives on this subject.

India and we Indians are indeed an enigma in most contexts, and scientific and industrial innovation is no exception to that. I recall the famous probability professor Eugene Dynkin of Cornell University asking me once in 1980 to name some top Indian probability researchers. I proudly rolled out many names, but with the sole exception of the internationally acclaimed leader in quantum probability, Professor K.R. Parthasarathy of the Indian Statistical Institute, New Delhi, Dynkin kept knocking each one as being outside India and not fitting his definition. I am sure my fate would have been the same had I tried some other field.

In any field of science or technology, and especially in their commercialization, it is easy to name many Indians at the very top. But it is likely that a vast majority of them are outside India and are probably also citizens of another land. Flattering as their Indian origin may be to India, for India to solve its own unique challenges and to become fully developed, it is necessary to foster the emergence in India of a cadre of great innovators in science and technology endowed also with entrepreneurial capabilities.

Considerable positive change is happening in India today in the formation of science and technology based enterprises, but it is sporadic and not due to the evolution of any nationwide systematized ecology. But even there, most companies formed are mere copycats and not based on really new homegrown science or technology inventions. There is certainly a need to assess honestly the inhibitors of Indian innovation and their possible cures, and the need is urgent.

With accessibility for non-technical audiences in mind, I have consciously striven to eschew my natural academic tone and provided the discussion at a reasonably popular level. But as someone aptly noted, one can make many things clearer, but nothing clear. This is particularly true for the complexities associated with development and innovation. Specifically, this book is not aimed at academic Development Economists or other specialists and does not expound any theory. Yet, I have little doubt that an open minded academic in these areas will find in it some worthwhile hypotheses to pursue for research.

No author can claim to have the last word on the topics discussed, nor can one expect universal acceptance of one's assertions. If this book would kindle even a modest discussion towards the evolution of some sound policies to accelerate the engine of innovation by Indians in India for India, then I would feel that a small fraction of my debt to my homeland has been repaid. My goal is to reach as many young and aspiring Indians as well as Indian policy makers as possible and goad them into prioritizing commercializable innovation both at the individual and at the macro levels for the betterment of the nation.

I offer this book humbly and with deep-felt respect for all the scientists and engineers who have made and are making significant contributions in India despite everything. I salute them.

Although it is the desire of every scientist to make his or her writing an everlasting legacy, nothing would please me more than to see this book become irrelevant someday by India asserting itself in a large way as a world leader in the creation and commercialization of science and technology.

**V. Ramaswami**
New Jersey, USA

# Acknowledgments

Writing a book requires the help and understanding of many, and it is my duty and pleasure to thank some key people.

Soundaram, my wife, in addition to bearing with my almost total absorption in this project that made her lug it alone in countless domestic and family matters, read several drafts and suggested many improvements both in style and substance amidst an avalanche of work related to the dissertations of more than half a dozen simultaneously graduating doctoral students mentored by her at Kean University. Several parts of the book were also read by my son Prem and daughter-in-law Shruti who, in addition to giving me the benefit of their fine training, respectively from the Harvard Business School and the Fletcher School of Tufts University, suggested some important references for study and inclusion. Dr. Ravi Subramanyam, a distinguished researcher and executive at Colgate-Palmolive, served as a constant sounding board giving great encouragement all through. But none of them spurred me more than my six-year old grandson Kiran and his four-year old sister Sahana who started writing their own books and finished them much before me.

I am extremely grateful for the encouragement and support from the following: Professor Anurag Kumar, Director, IISc., Padma Bhushan Dr. V. Rajaraman, Supercomputer Education Center, IISc., assessment expert par excellence Professor Elaine Walker of Seton Hall University, Indian diaspora expert Professor Nilufer Bharucha of the Center for Advanced Studies in India (CASII), noted English poet Dr. Sridhar Rajeswaran, Mr. P.C. Ramakrishna famous for his sonorous and unique voice matched only by his cultural erudition, and the great inventor Dr. Kicha Ganapathy of ColorEyeQ. My special thanks go to members of Spatika Digital Solutions for the quality promotional video on the book. I thank Mr. S. Raamassubramanian for some sound advice and my friends Ravi and Kritika Iyer, Mrs. Jaya R. Moorthy, and Dr. Suresh Singh for reading some specific chapters and commenting on them.

Mr. S. Kumaar of Gnanodaya Press, Chennai, handled all aspects of the production of the book in its various embodiments. The final product speaks for the quality of this longstanding institution of Chennai. I am fortunate to have had the help of Mr. S. Ravi Shankar - an artist recognized by Lalit Kala Academy - with the layout, and the beautiful art work that adorns the wrapper.

I shall remain eternally grateful to Ms. Usha Viswanathan who edited the chapters and made countless suggestions that not only improved my presentation enormously, but also increased my respect for those with formal training in English and the art of writing.

Although I have benefited much from many, the opinions and views expressed in this book are entirely mine as are any errors of omission and commission. I could, however, not have pulled off this significant foray of mine into a non-mathematical subject without the generous help of so many.

**V. Ramaswami**

# Dedication

The real inspiration for this book is a forty-five minute face-to-face conversation I was privileged to have with Dr. Abdul Kalam at Anna University, Chennai, just prior to his election as the President of India. His vision for India was grand, and his belief in the country and its young was unquestionable. He said that each one of us who has had some unique opportunities and experiences should share our ideas with India's young. I dedicate this book to the true jewel of India, Bharat Ratna Avul Pakir Jainulabdeen Abdul Kalam, who exemplified what India and Indians can achieve once they set their minds to it.

Dr. A.P.J. Abdul Kalam
Professor, Technology & Societal Transformation
**Anna University**
Chennai 600 025

Phone : +91 - 44 - 230 0886 • Fax : +91 - 44 - 230 0756
Email : apjabdulkalam@yahoo.com • apjabdulkalam@annauniv.edu

# TABLE OF CONTENTS

# Chapter 1

# The Dire Need

India's aspiration is to be in the league of developed nations soon, and the clarion call of its eminent scientists led by Dr. Abdul Kalam in the famous book *India 2020* [1] has given an added boost to it. A former National Security Advisor of India, Brajesh Mishra declared once that India intends to leapfrog into an advanced state of science and technology and does not wish to achieve it slowly and incrementally. It is in these contexts that we find the main subject matter of this book - applied research, innovation, and commercialization of research - relevant and important.

## India Now

Yogi Berra of American baseball fame, who is also known for his wit, is supposed to have quipped, "You've got to be very careful if you don't know where you are going because you might not get there." In the case of India, even knowing where it wants to go, it has to tread, nay sprint, with focus and resolve, starting with a good understanding of the challenges of the journey. Thus, there is no better place to start than with at least a quick assessment of the distance it has to traverse.

As of July 2015, it is estimated that India has a population of approximately 1.252 billion (about 125 *crores*). The data gleaned from *The World Factbook* [2] of the CIA shows quite a few interesting facts about India and the major challenges facing this very large nation even after sixty-eight years of independence. One may quibble about specific numbers, but they collectively do present a stark view even if we were to grant a large margin of error in the reported data.

India's GDP is the third largest in the world, coming behind only those of China and the USA. Yet, the per capita 2014 GDP of India, estimated at $5,900, is abysmally low when compared to those of developed economies (Germany $45,100, UK $39,500, USA $54,600). The wealth distribution in the country is so skewed that, according to a report in *The Hindu* [3], the richest 10% own nearly 75% of the wealth of the entire nation and have cornered 370 times as much wealth as owned by the poorest 10%. Not only that, 22% of the Indian population actually lives below the poverty line (*The Hindu* [3]) .

Going back to the CIA report [2], 67.3% of the Indian population is rural, suffers high unemployment, and is dependent for most part on agriculture wherein productivity is very low whether it is measured per acre of land or per person employed. In India, a sizable fraction of the GDP (17.9%) comes from agriculture, the comparable figures being respectively 0.6% and 1.6% for the UK and USA. The impact on the Indian economy of inefficiencies in the agricultural sector is, therefore, high.

Sanitary facilities are available to only an estimated 28.5% of the rural population, and to only 39.6% overall across the nation. Unemployment runs at about 10.7%. The country's health services are also meager as exemplified, for instance, in the availability of only 0.7 hospital beds per 1000 persons; the comparable figures for the UK and USA are 2.81 and 2.45 respectively. Overall, India's health expenditure in 2013 stood at 4% of the GDP while that of the US was at 17.1%, of Germany at 11.3%, and of UK at 9.1%. Even without factoring the lower per capita GDP of India, one sees that the outlays can barely meet the needs of the country which has increasing levels of chronic diseases like hypertension and diabetes, and frequent occurrences of various infectious diseases like dengue. The country may be one of the world's largest industrial powers and poised to become even larger through Foreign Direct Investment, but it has to make do "with an essentially illiterate, unhealthy work force" (Amartya Sen [4]).

Tables 1.1 & 1.2 at the end of the chapter, based on data given in *The World Fact Book* [2], give a side-by-side comparison of India with some developed countries on a diverse set of metrics. (The 2014 numbers being estimates are constantly under revision although the comparisons given here would remain valid at large.) Even a cursory examination of these tables will show that India has a long way to go before it can call itself developed. India's total GDP of 7.4 trillion dollars, reported in Purchasing Power Parity (PPP) terms, places it below only the US and China and above all other nations of the world. But India has to support such a large population (1.25 billion) that its per capita GDP is very low, and that population is growing at a much faster rate relative to developed nations. We note from Table 1.1 that almost half of the labor force in India is in the agricultural sector that also contributes a very large share of the GDP when compared to developed nations. India is not yet a big trading nation if one makes a comparison of imports and exports once again on a per capita basis. Although life expectancy at birth has improved significantly from 32 years at the time of India's independence, the number 68.13 quoted in Table 1.2 as well as the rates of infant and maternal mortality place India far behind the developed world.

In general, one needs to exercise caution in comparing the raw numbers reported in various tables. For instance, consider that India has nearly four times the population and less than half the GDP (again measured in favorable PPP terms) of the USA. These facts mean that the total expenditure of 4% of GDP spent by India on healthcare relative to the 17% spent by the USA, in reality translates to a per capita health care expenditure less than 3% of what the US spends per capita in PPP dollar terms. Also, many numbers, like the reported 0.7 physicians per 1000 in Table 1.2, belie the major quality differences in medical care that exist not only between India and the developed world, but even within various parts of India. Although India has become a notable destination for medical tourism even from the developed world and can boast of having some of the finest hospitals and doctors, a significantly large section of India has to make do with poor quality of medical care or none at all. Indeed, on most metrics that measure quality of life for individuals, India ranks far below the developed nations of the world.

Indeed, one can continue the lament and even play various blame games going all the way back to the British rule of India, external machinations, or even official incompetence and corruption in various administrations of independent India. But it is more productive to concentrate on those challenges that can be addressed by the nation in the present and in the foreseeable future.

## Progressing Against Great Odds

Lest the reader should get a wrong impression of my placing up front the current plight of India, let me say that this book is not an exercise in India bashing. I know all too well that the achievements of India have not come easily. The country and its scientific establishment have had to achieve too much with too little. They have had to endure the obstacles of trade barriers, sanctions, and export control regimen intended not only to deter India's growing military power but also the power to compete economically. Internal efforts at major indigenous development have been thwarted many times by foreign governments who would eagerly offer (outdated) technology at low prices at the nick of time killing the support from the Indian government for indigenous production (Chengappa [5]). Thanks to a lingering legacy of colonialism in the form of anglophilia combined with the admiration for America and the West among the powerful elite, scientists and technologists in India have always had to fight an entrenched preference for western goods and services. They also suffer disproportionate ridicule in the media even for small failures that inevitably occur when big challenges are undertaken. Furthermore, much of the Indian media join the chorus of their foreign counterparts in the cry that a country so poorly placed like India should refrain from spending as much as it does on science and technology, and should, instead, concentrate on its poor masses.

I am proud to be an Indian and proud of India for many of its achievements. Its vibrant democracy, its success in defying all predictions of the starvation deaths of millions of Indians (like those that occurred several times under the mighty British who were less able to prevent and may have even abetted them – Mukerjee [6]), and its incredible achievements in pharmaceutical, nuclear, space, and super-computer technologies despite the obstacles posed by the West are only a sample of its accomplishments. Mangalyaan, the program that launched an orbiter to Mars successfully, at a per kilometer cost "less than that of a one kilometer auto-rickshaw ride in Ahmedabad" (*Times of India* [7]) attests to the scientific and engineering talent in the country. India has indeed evolved to be a beacon of hope and a worthy example for the entire developing world in many ways. Its transformation into a developed nation is, therefore, of great import not only to itself, but also globally to many other aspiring nations.

Yet, my exercise will not be one of sycophancy or jingoism. We have enough of that, and at best it is counterproductive making the nation and its laudable efforts for progress even less credible. It is, therefore, important that we examine the situation in a critical and candid manner.

## The Role of Innovation

Be that as it may, the stark reality for the nation is the challenge of catapulting itself saddled with so many problems into the developed world. Attaining a high level of domestic innovation and milking its commercial benefits fully should assume a high priority in India's plans for development. But that will have to be done for India by India and will not happen by the benevolence or goodwill of any other power. "Make in India" alone won't do it if the process provides only Indian labor. The exhortation of Bhagavad Gita (Chapter 6, Verse 5) "*Uddharêt ātmanātmānam* – Elevate the self by thyself" is no less relevant in the context of innovation than in the context of spirituality.

A big shift to a technology and innovation based economy is not easy. It is certainly true that in the near term the massive employment of the largely unskilled people in the country can occur only in labor-intensive economic activities and industries (The Economic Survey 2014-2015, Government of India [8]). Granted also that capital constraints within the nation are a serious impediment. Hence, the drive to attract more foreign investment in India's manufacturing sectors and the need to launch the 'Make in India' campaign are necessary to address problems and may, indeed, be the panacea in the short run. But we need to keep in mind that manufacturing without innovation is no longer a sustainable option (See Zhong [9]).

> "Industrial latecomers now have to compete against China, whose massive, integrated machine has made it the world's factory floor and created a huge barrier to entry.... Bargain-price Chinese goods, produced at titanic scale, mean that even with India's [lower] factory labor costs ... manufacturers have to work harder to compete than they would have a decade ago." - Raymond Zhong [19].

The long run challenge, thus, cannot be met without accelerating the innovation engine of India significantly. Much as India may have leapfrogged in impressive ways in certain limited areas, there is a need for evolving at the national level a steady long haul plan of overall development, particularly with respect to commercializable innovation based on science and technology.

## Innovation Defined

What do I mean by innovation here? By innovation, I mean serious applied research and commercialization of such research resulting in new products and services that India can own and market globally, as well as any type of innovation that creates and transforms businesses in the country and provides yet another internationally marketable product which would enable India to increase its global capital base and capabilities.   It is in this focus that this book mainly differs from many of its peers.

The above is not to be confused with the mere lending of Indian bodies and even minds for work without any claims to the intellectual property they create. That enterprise has earned for itself the pejorative name 'body-shopping.' Much of India's software industry is built on that and is sucking the best of India's human capital away from Indian innovation and sustainable growth while bringing only limited benefits to India from a long-term perspective. The relative and quick monetary success of that sector has, unfortunately, deluded some into a false notion of progress. There is little realization that unless it is replaced by significant Indian innovation moving Indian industry's role into a much higher stratum, the arbitrage game of labor pricing that underlies it will come to an inevitable grinding halt as customers flock to cheaper pastures in other countries. This challenge has increased already due to the protectionist shrill cries that occur at increasingly louder levels in the customer nations about outsourcing of jobs. Advances in cloud computing pose an even greater challenge.

> "Big companies are shifting toward cloud storage and off-the-shelf software, reducing the need for Indian technicians and coders to maintain in-house data centers and write bespoke programs." – Raymond Zhong [9].

## Challenge of Ownership

Let me, therefore, emphasize that the most important issue with respect to innovation to be addressed in India is the Indian ownership of Indian innovation, and the commercialization of such innovation to increase the wealth of India in a sustained and sustainable manner. Today, even the little innovation that occurs in India, including in several government or government subsidized institutions, is often lost to other countries in the form of joint collaboration, work for hire, or as scientific publications.  Indian innovation cannot be owned by India unless it is paid for and supported by India, and until and unless the Indian government and industry make large investments to generate and protect Indian intellectual property.

The challenge of *swaraj* for Nehru and his peers of a young independent India was to ensure that value got added to Indian physical raw materials in India, and what got exported were not the raw materials but finished goods with much added value.  They had to ensure that India did not become a mere market for others but a large producer of industrial goods as well.  Due to the vision they had and the laws they laid down for limiting foreign ownership, particularly of land, India is not in the same situation as many countries in, for example, South America. In those countries, foreign corporations own most of the land while the locals are hired as laborers to slog on them for only a subsistence level pay.  Costa Rica's banana farms are good examples of this phenomenon that Nehru would have rightly called 'neo-colonialism.' Even with such vision, for several decades, India had to manage only with a 'screw driver technology' that made, at best, a final assembly in its factories. It took several decades for India to graduate into a credible manufacturing country with its own portfolio of products. Much of the success of India in this is owed to the infrastructure investments of independent India in power, water, education, and the like.

Today, in the scientific and technical arena, India faces a much bigger challenge than in the Nehru era. Too much of the equivalent of a screw driver technology is operating in India, particularly in software and BPO services. Large numbers of Indians, who could be moved to higher levels of the food chain, are slogging obscenely long hours in the new sweatshops - the code cutting warehouses, BPOs, and call centers. When we compare the current state of affairs with the past, it appears as though human capital has replaced physical capital, but otherwise nothing much has changed. Overcoming the current challenge has some additional difficulties too in the form of fast obsolescence of technologies and protective regimes of egregious patents and patent trolls.

## A Consortium Approach

In meeting such a daunting challenge, the government, academia, and industry need to throw their collective weight and forge a similar consortium like the one that marked the great forward marches in Indian space and missile technologies. The goal of that effort should be a targeted accelerated program of scientific and technological innovations for global commercialization for India's economic prosperity. The big questions for India are who will bell that cat, how, and when.

I see a big stumbling block here that is even bigger than the risk tolerance of Indian industry and available capital. As everywhere else in the world, a certain snobbery exists in academic circles in India too that frowns upon applied research and innovation as defined above, especially when it is tied to notions of commerce and profit. But, somehow, I feel the intensity of that is much higher in India than elsewhere. In addition to the higher premium Indian academia places on approbation from foreign peers, it may very well be also due to the lack of exposure to possibilities and the way the current academic systems limit faculty and researchers from reaping the benefits of participating in extra-academic engagements.

Certainly, there are nobler constraints as well. The ethos of the country over millennia has been that knowledge should be free and unfettered. Finally, a reason may also be found in the fact that a scientist like me wants to leave a permanent legacy of ideas and theories as opposed to contributions to ephemeral technologies. After all, technology does change fast, rendering innovations obsolete quickly; a recent example is the wireless communications arena which has moved in a very short time from 2G to 3G, 3G to 4G, and onto LTE etc. In addition, the majority of us form a culture obsessed with the fear of *mitya* (impermanence).

## A false dichotomy

The type of work I advocate, however, does not have to happen at the expense of basic or fundamental research. It is just that those engaged in the latter should take an active role in furthering the former as well. Today's scientists are not like some of the self-funded natural philosophers of the oligarchical rich nobility of the past. Taxpayers support them directly or indirectly. It is not much to ask that they also be cognizant of the financial and commercial needs of the nation and

chip in with their unique talent and skills that they would have acquired with much taxpayer subsidy.

My own personal experience as a researcher – I would consider at
least 70% of my work to be theory and algorithms of an academic type
– has been one negating the false notion that research cannot co-exist
with commercialization. On the contrary, I have found that while
commercial efforts become empowered through the involvement of
researchers, product development and commercialization in their turn
do serve as powerful catalysts for research by posing new questions.
This interplay can only increase the enthusiasm, output, and self worth
of the researcher.

> "Science is inherently open-ended and exploratory. Develop-
> ment is a closed loop .... For each time scientists come up with
> a thoroughly researched and fully comprehended solution,
> engineers show them yet another lumineu, yet one more pos-
> sibility." – Abdul Kalam [10].

A society may be strong scientifically, but if it does not complete the
cycle of the knowledge quest by bringing science to the development of
products and services, it stands to lose in the long run. No one can
champion a theory or concept into a product better than those who
created it, for their passions are not easily transferable even if their
knowledge is. Also, individual optimization by scientists to satisfy their
personal curiosities and to gain fame should yield, at least to a reasonable extent, to social optimization for one's country and the betterment of life for all. Obviously, applied success is also in the greater
interest of science in that the former can increase the ability to garner
more resources for the latter based on demonstrable payoffs. This is so
true even in fields like mathematics at first sight far removed from
engineering. Indeed, the famous mathematician Courant pointed out,
"To pursue mathematical analysis while at the same time turning one's

back on its applications ... is to condemn it to hopeless atrophy." Would any true passionate scientist want his or her field to be relegated thus?

## The Wealth of Nations

We have come a long way from Adam Smith and his days of sweat and blood that were the main tools to create wealth. Much of the wealth of nations today is created through research, productizing of research, and innovation in everything from industrial to business processes. How engaged are we in India in creating and investing in these as a nation? Are we grooming an adequate pool of inventors in India who will invent in India for India? Do we support them to succeed in the market place? Obviously we are not, since droves of them have left, are leaving, and succeeding elsewhere. And counterexamples related to commercialization of scientific innovations that can be shown are so few in number to be worth even a mention.

It is a sad fact that despite bragging to be the software capital of the world, India has not churned out one new technology business comparable to any of the new age giants of information technology and Internet commerce. When it is inconvenient, we should not claim it unfair to be compared with the developed world, and especially so if our honest goal is not incremental but to join the developed world. It is good to also bear in mind that many of the new giant oaks of the modern information age, despite their being in developed countries, grew from doubtful tiny acorns amidst much adversity similar to that currently encountered by start-ups in India.

The situation in India is so painful that even the celebrated doyen of Indian body-shopping chose to lash out at a premier Indian institution making the latter and its cohorts a scapegoat [11] for what is a national malaise of a much larger magnitude. True it may be that he and his ilk abet the continuance of the situation by going after quick money and not encouraging adequately the right type of Indian innovation. But shooting the messenger [12] does not by itself negate the bad news that perhaps raised his late concern and a sub-optimal expression of it.

Today, nations advance economically and wield their power not only through military might, but even more through the control of intellectual property and innovation based businesses that yield high margins and profits for their commercial enterprises. The information and pharmaceutical industries of the US are solid examples of the types of businesses that make a nation rich and make it possible for the US to offer its citizens a high standard of living. And, of course, there is also that industry of defense products that reaps some of the highest profits through constant innovation of weaponry that renders previous versions obsolete ever more quickly.

Small is no longer beautiful. Small and medium businesses are gobbled up by large multinationals that control much of the world's resources including innovative talent. Backed by their governments that preach globalization and fair trade mainly for others, they stem even the appearance of competition at all levels. This is an environment in which the stakes are much higher for a country like India that aspires to become developed, but is poised dangerously low on the hill of scientific innovation of the type resulting in much commercial value.

## National Security

Even a nation's security and ability to defend itself depends highly on its ability to innovate.

> "In today's world, technological backwardness leads to subjugation. Can we allow our freedom to be compromised on this

account? ... Should we not uphold the mandate bequeathed to us by our forefathers who fought for the liberation of our country of imperialism? Only when we are technologically self-reliant will we able to fulfill their dream." – Abdul Kalam [10].

A big price is being paid by India for its failure to support Indian innovators in India. As an example, consider the following.

"Ironically, Indian soldiers who fought against the Pakistan Army in the frozen heights of Kargil were equipped with obsolete field telephones, while the Pakistanis had sophisticated satellite phones based on technology developed by Indian software engineers and programmers in the West." – Stephen P. Cohen [13], p.106.

India is still too dependent on imports for its defense technology even as Indian born engineers and scientists are making better arsenals for other countries.

There is an even larger problem. In the world transformed by the information revolution, those condemned to depend on other countries for computer and communications equipment, systems, and networks are highly compromised in the area of their own security. In a recent conference in India, I was shocked to hear a learned speaker assert that networked banking should be totally secure in India 'because we use *ssa* protocols and encryptions' entirely oblivious to the fact that foreign corporations that license these technologies are obligated by their national law not to sell any secure technology their own governments cannot disable or break into. National security, today, has to contend not only with petty criminals and tactical theater issues, but also with the machinations of powerful nation states in the area of cyber security. If the next wars are to be waged and won, or better still averted, on the basis of information and cyber security, then innovation within India is absolutely essential for India's national security in many diverse fields including cyber security and communications.

## Economic Neo-colonialism

Without significant innovation in India, Indians are asking to be subjected increasingly to a new type of economic colonialism. They shall continue to remain at the lower ends of the food chain, their hard work enriching other nations even more while bringing but a semblance of relative improvement in their own plight. Yes, at an individual level and in comparison to one's own past, many may find themselves better off, but not as a collective and as a nation. The grandchildren of those enjoying the fleeting prosperity of today may be placed back where their grandparents started.

## The Urgency

A spurt in innovation does not happen overnight. To make a difference at the macro level, it has to be nurtured consciously over the long haul. It takes a special type of national mindset and environment. Even when it happens, we need the support of capital and business to reap its benefits fully. Delaying a serious commitment to accelerate innovation and to erect support structures for innovation in the country can only widen the gap even more and make it much harder to catch up than it is already. The need of the hour is the careful evolution of policies and a sagacious allocation of resources.

I hope even this brief discussion has convinced you of the dire need to accelerate innovation in India. We will in Chapter 2 take up a general discussion of some factors that are conducive to an innovative climate and those that inhibit innovation. The rest of the book will be an assessment of each of them in the Indian context and an examination of some possible avenues for improvement.

**Table 1.1: Comparison of India with Some Developed Nations**
Economic Indicators – 2014 estimates
Source: The World Factbook, CIA

| INDICATOR | INDIA | UK | GERMANY | USA |
|---|---|---|---|---|
| Population in millions | 1,251 | 64 | 81 | 321 |
| **GDP & Savings:** | | | | |
| Total GDP in billions of US $ | 7,277 | 2,435 | 3,621 | 17,470 |
| Per capita GDP in US $ | 5,900 | 37,700 | 44,700 | 54,800 |
| Gross Savings as % of GDP | 30.1% | 10.8% | 26.3% | 17.3% |
| **GDP by sector:** | | | | |
| Agriculture | 17.9% | 0.6% | 0.9% | 1.6% |
| Industry | 24.2% | 20.6% | 30.8% | 20.7% |
| Services | 57.9% | 78.8% | 68.4% | 77.7% |
| Gross National Savings/GDP | 30% | 11.9% | 26.7% | 18.8% |
| **GDP Composition by Sector** | | | | |
| Agriculture | 17.9% | 0.6% | 0.7% | 1.6% |
| Industry | 24.2% | 19.9% | 30.4% | 20.6% |
| Services | 57.9% | 79.5% | 68.9% | 77.8% |
| **Labor Force by Occupation** | | | | |
| Agriculture | 49 % | 1.3% | 1.6% | 0.7% |
| Industry | 20% | 15.2% | 24.6% | 20.3% |
| Services | 31% | 83.5% | 73.8% | 79.1% |
| Below poverty line | 29.8% | 15% | 15.5% | 15.1% |
| Inflation rate/Consumer Price | 6% | 1.5% | 0.8% | 1.6% |
| Commercial bank prime rate | 10.3% | 4.45% | 2.47% | 3,25% |
| Exports (billion $) | 342.5 | 480.8 | 1,492 | 1,633 |
| Imports (billion $) | 508.1 | 479.7 | 2,387 | 1,592 |
| Unemployment | 8.6% | 6.2% | 5.0% | 6.2% |

**Table 1.2: Comparison of India with Some Developed Nations**
Other Indicators – 2014 estimates
Source: The World Factbook, CIA

| INDICATOR | INDIA | UK | GERMANY | USA |
|---|---|---|---|---|
| Population Growth Rate | 1.22% | 0.54% | -0.17% | 0.78% |
| Death rate per 1000 | 7.32 | 9.35 | 11.42 | 8.15 |
| Infant mortality per 1000 live | 41.81 | 4.38 | 3.43 | 5.87 |
| Maternal mortality/100,000 live births | 174 | 9 | 6 | 14 |
| Life Expectancy at Birth | 68.13 | 80.54 | 80.57 | 79.68 |
| Dependency Ratio | 52.4% | 55.1% | 51.8% | 50.9% |
| Health Expenditure/GDP | 4% | 9.1% | 11.3% | 17.1% |
| Physician density per 1000 | 0.7 | 2.81 | 3.89 | 2.45 |
| Hospital beds per 1000 | 0.7 | 2.9 | 8.2 | 2.90 |
| Improved sanitary facility | 39.6% | 99.2% | 99.2% | 100% |
| Education expense/GDP | 3.8% | 6% | 5% | 5.2% |
| Landline phones per 100 | 2 | 52 | 58 | 41 |
| Mobile cellular per 100 | 76 | 123 | 123 | 100 |

## Notes & References

[1]  A.P.J. Abdul Kalam & Y.S. Rajan: *India 2020, A Vision for the New Millennium*, Penguin Books India, New Delhi, 1998.

[2]  The World Factbook, CIA, https://www.cia.gov/library/publications/the-world-factbook/geos/in.html

[3]  "India's Staggering Wealth Gap in Five Charts," *The Hindu*, December 8, 2014, http://www.thehindu.com/data/indias-staggering-wealth-gap-in-five-charts/article6672115.ece

[4]  Amartya Sen: "India trying to be industrial giant with an illiterate, unhealthy labor force," *MSN News*, September 7, 2015. http://www.msn.com/en-in/news/national/india-trying-to-be-industrial-giant-with-an-illiterate-unhealthy-labour-force-amartya-sen/ar-AAcK1Pp

[5]  Raj Chengappa, *Weapons of Peace*, Harpercollins Pub. India Pvt. Ltd., 2000.

[6]  Madhusree Mukerjee: *Churchill's Secret War: The British Empire and the Ravaging of India during World War II*, Basic Books, A member of the Perseus Book Group, USA, Paperback, 2010.

[7]  "India's Mangalyaan Ride Cheaper Than Auto, Cost Rs. 7 a KM: Modi," *Times of India*, September 29, 2014. http://timesofindia.indiatimes.com/india/Indias-Mangalyaan-ride-cheaper-than-auto-cost-Rs-7-a-km-Modi/articleshow/43779945.cms

[8]  *Economic Survey 2014-15*, Government of India, http://indiabudget.nic.in/vol1_survey.asp

[9]  Raymond Zhong: "Manufacturing Bust: Wealth-producing Factories Face Headwinds as Global Trade Slows and Consumers Age," *Wall Street Journal*, November 24, 2015.

[10]  A.P.J. Abdul Kalam (with Arun Tiwari): *Wings of Fire, An Autobiography*, Universities Press, 1999.

[11] "Has IISc. Contributed to Society, Asks Narayana Murthy," *The Hindu*, July 16, 2015. http://www.thehindu.com/news/national/has-iisc-contributed-to-society-narayana-murthy/article7426651.ece

[12] "Bengaluru: City Scientists Decry NR Narayana Murthy's Remarks on Poor Science Research," *Deccan Chronicle*, July 17, 2015. http://www.deccanchronicle.com/150717/technology-science-and-trends/article/bengaluru-city-scientists-decry-nr-narayana-murthy's

[13] Stephen P. Cohen: *India: Emerging Power*, Brookings Institution Press, Washington, DC, 2001.

## Chapter 2

# Factors Impacting Innovation

There are many factors that impact the innovation capability and the innovation success of a nation. The goal of this chapter is to list some important ones and to examine them at a general level. In later chapters we shall delve deeper into each one of them, with particular reference to the Indian context. For reasons of space, we are constrained to limit our discussion to only some of the most important factors.

## 2.1 Society and Culture

Even in highly developed societies, major innovators and wealth creators form only a minuscule minority. Nations that can identify and groom them have, indeed, a special advantage. Some nations are consistently better not only at innovating but also in generating and maintaining a constant pool of innovators. Something about their social and cultural makeup enables that, and that is worth studying. And, of course, there is no better place than America for that exercise. Both indigenously and through immigration, the USA has remained the true citadel for innovation for many decades.

Within the factor construct of 'Society and Culture' in the context of innovation, I like to include the following: the vision and aspirations of the people - the very definition of what it means to have "arrived"; the level of collective discontent towards status quo; readiness to challenge tradition and willingness to embrace change; the level of independence, initiative, and accountability; a long haul perspective; willingness to take risks, and ability to tolerate risks. Let us examine them in some detail.

## Noble Discontent

I believe that a noble discontent with status quo and the burning desire to become wealthy are the foremost drivers of innovation. Both these are hallmarks of America. Here is what one of India's great technologists has to say:

> "My impression of the American people can be summarized by a quotation from Benjamin Franklin, 'Those things that hurt instruct!' I realised that people in this part of the world meet their problems head on. They attempt to get out of them rather than suffer them." – Abdul Kalam [1].

As a country, America is blessed with an abundance of people who refuse to put up with inefficiencies, reject hard and repetitive work, constantly search for improvements, and are unwilling to endure avoidable or surmountable inconvenience even when its removal is challenging or risky.

The nobility of the discontent alluded to above manifests as a creative and constructive response to eliminate the discontent through some innovation as opposed to meek accommodation or counterproductive reactions. A large pool of persons with such noble discontent can be expected to innovate much. In this context, it is worth noting that when manual labor is very cheap, there is no great incentive to innovate, for pain is shifted away to someone else quite easily. One may argue that offshoring of manufacturing is such a shift by developed nations at a macro level; its long-term consequences on developed societies are only beginning to emerge.

I am constantly amazed by the plethora of new products that are designed and brought to market each year in the USA (made in China, of course!), most of which resolve but some simple problems. An example is a simple zero-technology device for taking out the seeds of a pomegranate efficiently; see Figure 1.1. The device does not have even a single moving part or a screw. Such products sell in the hundreds of thousands bringing high rewards to the inventor and to the business that seizes the opportunity to market them. What makes it possible is an efficient network of sales channels, and legal protections that fend off copycats. The real engine for growth in America is indeed the small business, and Americans start them in droves even though the risk is high and many do fail [2].

Figure 1.1: A simple device to take out pomegranate seeds

Yet there are also cultures in the world where most people have tremendous tolerance and fortitude and will perpetuate the use of the same inefficient processes and tools across many generations in sheer apathy or in deference to tradition and embrace change only slowly and reluctantly even when it comes. My pet example is how, despite the best of mechanical engineers in India, the bucket pump so essential to many for even drinking water there has remained essentially unaltered for decades and decades. Whatever may be the underlying pathology behind this, it goes without saying that such cultures do not usually exhibit a high innovative spirit in technology commercialization.

It is gratifying to see in India an increasing level of innovation leading to new and improved products in the small business sector. But it seems to occur at a much lower rate, and the ones that occur do not get comparable support to become big successes. We will discuss some remedies in later chapters.

The spirit of innovation needs to be nurtured from a young age, both in the home and in schools, through education that emphasizes problem solving and projects that elicit creativity. Society should also find effective ways of recognizing and rewarding innovators at all levels in a visible manner and cheering them on.

## Attitudes to Wealth

As noted, the other driver that is most important is the burning desire to become wealthy. Just becoming better than what one was or what one's near and dear are would not suffice. The vision has to be much larger to drive one to create and innovate and to willingly take on the challenge of bringing one's ideas and products to the market place.

People should be encouraged to consider the ability to earn wealth as a talent as important as any other. Wealth is needed even to do good to others, and those who can earn it should consider it a social duty to earn as much as they can, so that they can help others in big ways. Directly as outright charity or indirectly through the provision of jobs, pursuits of wealth do contribute to human welfare. What makes wealth sinful is not how much one earns or wants to earn, but how it is earned and on what it is spent.

Cultures in which a substantial section of the intelligentsia touts the virtue of contentment beyond limit and approaches wealth with disdain and suspicion are not the ones where one will find much innovation and entrepreneurialism. Such cultures also corrupt their young by instilling an attitude of aversion towards even legitimate material pursuits, thereby giving a convenient alibi for not even trying. A recent book (Grubman [3]) delineates the ill effects of wealth aversion on individuals and families, and it is fair to hypothesize that there are similar effects at the macro level too.

As a practicing Hindu with a deep interest in Vedanta, I aver that balance and contentment are absolutely important, and that one should never compromise one's higher values in the pursuit of wealth. But that is not to say that one should not pursue wealth. We do live in the mundane as participants in a society where so many are less fortunate and can be helped through pursuits of wealth that can create jobs and support charity. Even the great ascetic gurus of Hinduism can be claimed to have agreed with this perspective (see Adi Sankara [4]). Also, why else would Hindu scriptures put wealth and security (*artha*) as the second most important value to a person right after righteousness (*dharma*)?

**Personal Traits**

A high level of self-reliance and personal accountability fosters the spirit of innovation, as does the willingness to take risk. For innovators, it is less important to be always right than it is to be useful and successful. It is better to have tried and failed than not to have tried at all. Cultures that place a premium on being right and conforming to tradition are at a great disadvantage when it comes to innovation.

Unlike research, the process of innovation and its commercialization involves much teamwork. This is an instance where extroverts may even have an edge over introverts. Societies that do not provide adequate opportunities to the young by way of team activities in the form of sports and team projects may also limit their future success potential. So is the case with societies where the burden of tests and test preparation on children consumes so much of their time that little is left for creative forays.

## Protection of Intellectual Property

Even in the developed world, inventors and their employers lose many commercial opportunities to ego gratification efforts, not mindful of the collateral. These occur in the form of research publications or public dissemination of work without them being streamlined properly through internal reviews for intellectual property protection. Recently, there is also the growing insidious practice of large corporations getting significant innovation and research for a paltry amount by crowdsourcing through highly advertised competitions. These stroke the ego of a large number of enthusiasts and purportedly encourage science and technology, but many times only to get for the cheap what may otherwise entail a significant R&D investment. This is something for India to pay watchful attention to and regulate so that no individual is fooled into giving a carte blanche without being properly compensated for the innovation.

India needs to make a careful study of how much intellectual property is lost to foreign nations through collaborations by Indians as post-docs and visiting research scholars, sometimes sent abroad for such assignments even at India's own expense. In legal contracts with much fine print, many such scholars sign away all rights to intellectual property even directly generated by them under the legal umbrella of 'work for hire.' Such giveaways by a person are at the expense of the society that has previously invested, either directly or indirectly, on that individual. India owes itself to make appropriate international agreements and contracts to safeguard against such losses.

## Level of Probity

Corruption is a great enemy of innovation. In the process of bringing an innovation to the market, innovators need to be able to find some whom they can trust. They also need to protect their intellectual property rights with patents and other legally enforceable contracts. In societies marked by low levels of probity, such protection and trust are hard to achieve. That, in turn, inhibits innovation drastically by increasing the risk to potential entrepreneurs. As a result, many are forced to either seek the safe harbors of paid employment or succumb to the option of premature sell-outs and end up with a much lower personal return. When the employer or buyer turns out to be foreign, an obvious loss to the nation is incurred as well.

## 2.2 The Economy and the Market

### Innovations – Large Scale and Consumer-centric

For small innovations aimed at consumers, it is essential to have a large addressable market of customers who can afford and are willing to spend money for improved or new products that yield greater comfort, enjoyment, or ease. Societies without such a consumer base cannot support a sustained level of innovation by providing a market for new products. Also, in societies that do not have an organized safety net of life and disability insurance, social security schemes, and pensions, even if the income is high, savings rates have to be necessarily high, and that reduces the amount of disposable income. Similarly, high tax rates tend to create much black money, most of which is rarely put into productive economic use.   In short, there are many macro economic variables that have a direct impact on the level of innovation even at the consumer level.

In addition, the presence of an effective distribution channel is equally important.  Developed societies have enterprises whose very business is the support of inventors in the process of securing their intellectual property and marketing their invention.  Some of these provide the needed financial support to the inventor by fronting various expenses involved in the legal and pre-sale processes. Various agencies set up by the governments to help small businesses provide encouragement, including through favorable loans and equity participation.

As for large-scale innovation, such as, for example a revolutionary drug for a disease or a new fighter plane, innovation, typically, occurs in the large industrial sector or the government.  The key drivers there are the overall risk tolerance by large companies and some macro factors like interest and inflation rates that drive project costs and payback time.  Some of these macro variables have a significant impact on businesses' willingness to take a long haul point of view and to invest in R&D leading to major innovations. Many industries, like defense, also depend on government outlays for research in general and for research outsourced to industry in particular.  The captive and supporting market provided by the government is a great catalyst for large-scale innovation.

## Market Competition

A highly competitive economy with a large number of suppliers and efficient distribution networks is a major driver of innovation. When businesses have to compete, they are forced to constantly upgrade their offerings and to innovate. Innovation is also driven by the desire to maintain a company's competitive moat [5] that gets eroded over time as competitors start making clones of their successful products or innovative business processes. They are also forced to innovate to diversify their product portfolio and, thereby, reduce risk. Efficient distribution channels bring new innovations quickly to the market, reducing investment risk and bringing rewards faster to implementers. As opposed to this, markets that are marked by less competition and that operate, essentially, as sellers' markets result in an entitlement system of assured profits, which removes the urgency or necessity for innovation.

Sometimes, even in advanced economies, pockets of serious market inefficiencies can flourish. Thus, in the USA, until 1984 such was indeed the situation in the communications industry with government having yielded for decades a regulated monopoly status to AT&T. Although much innovation did occur in AT&T Bell Labs (laboratory researchers even won eight Nobel awards), many innovations were never brought to the market. Due to an assured customer base, the company had little incentive to adapt to fast occurring changes all around. It had little incentive to commercialize its research due to rate caps on the profits it could earn and retain. As a result, the US lost much of its leadership in telecommunications and was able to gain some of it back only after the break up of AT&T and the creation of a more competitive environment, which led to benefiting the customer and the nation [6],[7].

## Hierarchy

Hierarchical societies have unwritten norms where decision making and thinking become the prerogative of only the top few. In such societies, most changes occur top down, and bottom up thinking is highly discouraged. With those feeling the pain mostly out of a constructive feedback loop, little incentive remains at the bottom to identify areas for improvement, let alone innovations.

## Gender Gap

Most countries lag significantly behind in innovation by women. This phenomenon is not restricted to the developing world. For example, only "7.5% of regular patent and 5.5% of commercial patent holders (in the USA) are female ... women's underrepresentation in engineering and in jobs involving development and design explain much of the gap" [8], [9]. India may have a specific advantage in this regard in the sense that there is a sizable representation of women in engineering education and also in engineering and design related jobs. Yet, it is reasonable to suspect that India is also very much in the same situation as the USA or worse when it comes to patents and intellectual property owned by women. Much of this phenomenon may be due to sociological causes that deserve great attention. Shutting out nearly half of humanity from an important activity like innovation does not serve the world at large. For women, it has caused major burdens even in developed societies like the USA [10],[11]. It behooves all to accelerate the participation of women in innovative endeavors in all fields.

## Multiplicity Effects

Innovation tends to breed on itself. High technology societies invent not only consumables and capital goods, but, even more importantly, invention tools accelerating innovation itself. A good example of this is 3-D printing [12], which enables fast prototyping at low cost and has, thereby, become a force multiplier for innovation.

## 2.3 Financial & Business Support

No idea is worth anything unless it is implemented. For innovations, financial support from the larger economy is the fuel that can launch them into the farther space of commercial success. The larger the potential market for the invention or larger the developmental effort, usually, the larger are the financial needs.

### Financial factors

Driving the success of innovation in the developed world are many financial support factors: investment by industry in research and development with a long haul perspective; risk tolerance and willingness to assume risk by industry and investors; entrenched culture of R&D in industry; availability of venture capital from venture capitalists, private investment, and funds; government sources like the Small Business Administration; loan and tax subsidies by government for business start-ups (used in the USA to uplift poor economic zones and to increase employment). Added recently to this milieu are some new and imaginative schemes like Internet based crowdsourcing to help smaller efforts.

### Industry R&D

Investment by industry in R&D is extremely important. The role of this factor can be seen from Table 2.1, which provides some data for the period 2006-2008 from the USA. Note the striking differences between those companies with no R&D activity and those with R&D activity.

In addition to enhancing the slate of product offerings, market opportunities, and profits for individual companies, the presence of well funded and well managed R&D programs in its companies creates

TABLE 2.1  Innovation incidence percentage - US 2006 -2008 - with & without R&D

| Company type | Companies (thousands) | New or significantly improved product | | | | | | New or significantly improved process | | | | | | | |
|---|---|---|---|---|---|---|---|---|---|---|---|---|---|---|---|
| | | Any good/service | | Goods | | Services | | Any process | | Manufacturing/production methods | | Logistics/delivery/distribution methods | | Support activities | |
| | | Yes | No | Yes | No | Yes | No | Yes | No | Yes | No | Yes | No | Yes | No |
| All companies | 1,545.1 | 9 | 86 | 5 | 91 | 7 | 88 | 9 | 86 | 4 | 91 | 3 | 92 | 7 | 88 |
| With R&D activity | 46.8 | 66 | 30 | 52 | 45 | 38 | 58 | 51 | 44 | 34 | 62 | 20 | 76 | 36 | 60 |
| < $10 million | 44.8 | 66 | 31 | 51 | 45 | 38 | 58 | 51 | 45 | 33 | 63 | 19 | 77 | 35 | 61 |
| ≥ $10 but < $50 million | 1.3 | 70 | 25 | 64 | 31 | 42 | 52 | 56 | 38 | 44 | 50 | 30 | 64 | 43 | 51 |
| ≥ $50 but < $100 million | 0.3 | 76 | 18 | 71 | 24 | 51 | 43 | 69 | 25 | 57 | 36 | 43 | 50 | 55 | 38 |
| ≥ $100 million | 0.4 | 81 | 15 | 77 | 19 | 56 | 37 | 71 | 22 | 60 | 34 | 54 | 40 | 61 | 32 |
| Without R&D activity | 1,498.3 | 7 | 88 | 3 | 92 | 6 | 89 | 8 | 87 | 3 | 92 | 3 | 93 | 6 | 89 |

NOTES:  Survey asked companies to identify innovations introduced in 2006–08. Sum of yes and no percentages may not add to 100% due to item non-response to some innovation question items. Figures are preliminary and may later be revised.
SOURCE:  National Science Foundation/Division of Science Resources Statistics, Business R&D and Innovation Survey, 2008.

See [13] for a more detailed discussion of the NSF report.

for the nation a climate that encourages invention at the macro level. It raises the vision and opportunities of many potential inventors for generations to come.

## Private Investment & Venture Capital

It is highly rare that an individual inventor or even a small group of inventors has high levels of business acumen or experience. Inventors are often technical people with specialized expertise and poor business aptitude. For them to succeed in the business world, they need to augment their technical creative skills by a variety of business skills. The process of taking one's invention to the market involves many steps, and each requires a diverse set of skills, starting all the way from the creation of a sound business plan to mundane aspects of budgeting and accounts and day-to-day management. In certain areas, there are also the minefields of regulation and patents to cross over and gorillas to fend off. Consumer products, particularly, incur significant up front costs related to marketing, advertising, and distribution. Crossing the dreaded Moore's market chasm [14] between the early adopters and the market majority takes much skill and money and often becomes the death trap for many start-ups. Innovations at larger scales need to be sold to large corporations or government. Often support from experienced, well-connected, and well-heeled investors becomes essential.

The literature on all the above aspects of the start-ups world is extant. For a comprehensive one-stop source, I have found the guide from Dartmouth Entrepreneurial Network [15] to be an excellent book. It discusses most of the important challenges along with prescriptions for meeting them effectively.

A well-developed ecosystem provides not only financial support, but also brings to the inventors a variety of services filling their gaps. Many venture capital firms have their own army of lawyers, accountants, and management personnel whose costs are shared across many ventures supported by them. The ventures themselves are cross-pollinated to increase their chance of success and payoffs.

There is an extensive literature on the important role of venture capital and other sorts of finance for start-ups. Chapter 12 of the Dartmouth guide [15] is a good place to get a general understanding. References [16] and [17] provide useful information on how things operate in the USA. While these may not be of terrible help to an entrepreneur in India, they certainly will be valuable reading for policy makers and those wanting to set up venture capital and private equity shops.

Thanks to the phenomenal success of Silicon Valley, we also have many books that chronicle specific tales of successful companies (and failures) written in a way to offer much learning and guidance for aspiring entrepreneurs. I have included two [18], [19]; the second one is, particularly, educational about the pitfalls to avoid. Reference [20] is a fun read. (Does your engineering school's library have copies of such books?)

In the USA, for an increasing number of youngsters, and, particularly, for those of Indian origin, having a start-up venture of their own has become their dream. Many are able to realize it, thanks to a working spouse who can serve in their home finance portfolio as the "income" part balancing the risk of the "equity" part of the entrepreneur. To some degree the safety net provided by the parents who have done well in the USA also plays a role in encouraging these adventures - at least as a psychological backdrop if not directly in the form of financial assistance.

Just encouraging the young to delve into some of the literature on entrepreneurship alone would light up a fire in them. My own evolution from a theoretical mathematician (that I continue to be) and a paid employee to an adventurer in this realm is due to such forays, ignoring occasional comments from well meaning friends in research that I may be "walking over to the darker side."

## 2.4 Educational System

A country aspiring to have many great innovators needs an educational system at all levels geared to generating them. Innovation demands much more than systems aiding only the assimilation of knowledge or scientific facts, but an entire ecology focused on self-learning, exploration, problem solving, and thinking up the future.

### Instructional Methods

Systems, where instruction is unidirectional and focused mainly on assimilation and fostering an ability to repeat and mimic, are not conducive to the generation of many innovators. Every science and engineering instructor must read the experiences in Brazil narrated by Richard Feynman [21]. Here are some excerpts.

> "I figured out that the students had memorized everything, but they didn't know what anything meant ... When they heard 'light reflected from a medium with an index' they didn't know that it was material such as *water*. They didn't know the 'direction of light' is the direction in which you *see* something when you are looking at it and so on. Everything was entirely memorized, yet nothing had been translated into meaningful words." - Richard Feynman [21]

Professor Feynman found that the Brazilian students could recite physical laws with fantastic precision but could not answer the simplest questions that involved an application of the very same laws.

My own professor, Marcel Neuts, a famous applied probabilist who has also made notable contributions of value to the practice of that subject, once riled up many in a US conference with the comment (reflecting also his knowledge of classical English poetry), "The sad fact is that the vast majority of queueing theorists have had no interactions whatsoever with a real queue except as those *who stand and wait.*" So, some of the complaints above are more widely applicable to many fields and environments.

Given Brazil's great strides forward, I would assume that Brazil has come a long way from the description of Feynman, but my own observation is that the bulk of the educational system in the bulk of the developing world, including India, has stagnated badly and needs a thorough overhaul. That overhaul cannot be made by bureaucrats and politicians, but only by scientists and educators of high caliber. Unfortunately, in many of these countries, these latter groups are least empowered, or are too elitist to get their hands dirty by stepping out of their ivory towers.

The problem is, certainly, the legacy of a colonial system that condescendingly gave the colonies a hierarchical "I tell you so" kind of an educational system geared primarily to rearing bureaucrats, clerks, and humble servants and not masters of their own destiny. (Sorry, Sir George Otto Trevelyan, your tome [22] didn't convince me otherwise about your uncle's contribution!) India has managed to change that to some notable degree, thanks to visionaries like Tata, Nehru, Homi Bhabha, and Mahalanobis, through the creation of many fine higher technical institutes like TIFR, the IITs, and ISI that quickly gained high stature as IISc, India's grand pioneer. These set the foundation for the later flowering of research and engineering institutions like BARC and VSSC, by visionaries like Bhabha and Sarabhai.

The bulk of the educational system in India, however, has remained where it was, and any progress has been minuscule. Combined with political corruption and vote getting counter-productive cries for teaching even science and engineering in local languages (at a time when being global is the way to economic salvation), democracy itself seems to get in the way of some countries like India. Is it surprising that we have a lot of catching up to do in the area of innovation and applied research for commercialization? If one may paraphrase the famous British writer Aldous Huxley who said a similar thing about politics, one must say that the shock is not that there are not many globally recognizable commercial innovators in these places, but that there are at least a few who, despite it all, do shine and thrive as inventors and entrepreneurs, albeit, in smaller degrees.

## The Sociology

Sociology plays a role in education too. The more hierarchical the relationship between the teacher and the taught, the greater is the disconnect between the two. Innovative talent grows in an environment where, in technical matters, one can question and challenge fearlessly and unmindful of rank, and where such challenges are not misconstrued as lack of respect. The best teachers are those who are clearly cognizant of the reality that new discoveries will replace even the most admired theories of today, and, therefore, seek out students who will challenge and question and not those who just take in instruction as silent spectators.

## Integration of Theory & Practice

The larger the disconnect between theory and practice in educational institutions, the less is innovation in the country. The less the presence of faculty who have applied their knowledge in the practical world, the less is the innovation. In developed nations, these gaps get bridged in many ways. A specific example is the presence of industrial laboratories in universities. They offer students internship training that gives them practical experience. Technologists and researchers from industry augment the faculty as valuable adjunct instructors. Much stimulus to creative activities of both faculty and students occurs through the presence of such laboratories in close proximity. Many research activities in engineering and the sciences start with a real problem and need as they indeed should, and do not end up as insignificant incremental advances of no apparent practical or commercial value. Also, the entire educational system permits a free flow of exchange between university and industry to the betterment of both. Elitism in academia that looks down upon industrial counterparts is a killer. So is any attitude in industry that brushes aside informed faculty as mere paper tigers. The stronger the university-industry partnership is, the more the innovation.

## Silo Based Systems

One other issue related to innovation and entrepreneurship is the breaking up of the old silos of compartmentalized education that confine students to narrow specialties. In places like the USA, students take courses from multiple disciplines and departments enriching themselves to apply their specialty to a diverse set of fields. In areas like statistics that provide generic tools, the absence of exposure to current fields of engineering application is seen to be disastrous, for such education is considered as only yielding a bunch of specialists that are like carpenters who have learned to make perfect joints, but cannot put together even a small piece of furniture. This type of a realization has forced universities in developed countries to encourage inter-departmental and inter-disciplinary efforts at all levels. The good statistics departments I know in the USA have set up consulting services for other university departments and local industry and agriculture with significant payoff. These help to enlarge the domain knowledge of students in key application areas and hone even research activities to relevant needs.

The better engineering schools equip their technical students with business acumen and entrepreneurial skills. Entrepreneurship programs at places like the University of Rochester and Stanford are fine examples. The (free) Coursera course (*http://www.coursera.com*) on technology commercialization by Mark Wilson of the University of Rochester is one such example. The profound interview embedded in this course with Dean Clark and another one with Duncan Moore, Vice Provost of Entrepreneurship, alone would give an educational leader many ideas worth pursuing. The entire program is geared to enable engineering students to start high-tech start-up companies. Many departments of engineering and management jointly support this effort, working together cohesively to train students, holistically, both in theory and practice. Those programs already have many significant success stories in the tough world of business. An example is a new course on Product Management at Harvard Business School that immediately obtained the support of practitioners and has since generated several successful start-ups by its students (See Reference [23]).

The above types of efforts can occur at an educational institution without the blessing of a big government. Making these efforts a reality requires no more than modest resources, a vision, and a good plan of implementation. Shedding some academic arrogance and taking in partners from industry of the type "Been there, done it" is also needed.

## Technology Incubators

A new trend is to have technology incubators right inside campuses. These allow participation by faculty and students for sweat equity. The university also gains through both equity participation in the companies formed and a share in intellectual property that is generated. The private universities in the US have found this to be a lucrative area and are growing these types of activities across the board.

## Elitism & Entitlement

Elitism in education and research that looks down upon engineering and applications is not helpful. Certainly, we need scientists engaged in basic and theoretical research who are left alone to do just that. But the engagement in such activities, except perhaps for a very small number, should be earned as a privilege and not obtained as lifetime entitlement. Even that should be based on future potential and not past glory. In general, while a long rope should be given to a promising beginner, others should be asked to earn it repeatedly through demonstrable achievements confirmed by processes of evaluation. The evaluations should not only judge effort and final results, but should calibrate success or failure against the level of challenges undertaken and the risks and potential payoffs associated with them. Best practices in this are what gave Bell Labs its eminence as an unparalleled research institution. Subsequent abandonment of them is what accounts for its current emaciated state.

The dream of an Indian scientist may be to win a Nobel prize or international (understood mostly as Western and particularly US) approbation for oneself, but that dream has to accommodate the needs and priorities of those footing the bill and the essential role of a scientist as a contributor to society through specialized knowledge. Most of us in the US earn the privilege to be a researcher year after year and are better for it. Similarly, Indian industry, which is making major philanthropic gifts globally for higher education and research, should equally support Indian innovation through awards and endowments *in India.* Visible awards in India should cover applied research as much as the theoretical and esoteric and should not be restricted only to academia. Having a balance of people in the selection committees with the representation of government based scientific institutions and industrial R&D may be helpful as also the institution of special prizes for this category separately.

## The Role of Merit

While primary and even bachelor's education could be the right of everyone, still higher education must be a privilege earned through academic merit only. Reservations in the upper echelons of education are an absurd concept, particularly in a country with limited resources where the cost of mediocrity is high. The right way to remedy underrepresentation of various groups is to seek out and identify the brightest among them and to groom them early on with additional support, both economic and academic. Special recognition should be given to mentors who conscientiously undertake such efforts. Similarly it is also absurd to have a "one shoe fits all" approach – even to high school education. Students should be channelized into programs by their ability and aptitude. A major strength of the best educational institutions in highly innovative societies is the meritocracy that prevails in them. An educational system that is geared only to lifting up the masses can make the terrible mistake of snuffing creative talent and robbing the nation of many potential major advances.

In any walk of human activity, the bulk of progress occurs due to a very minuscule minority. Not to identify, encourage, and nurture the most promising in that select group is a travesty. Good public school systems in the US, particularly those in New Jersey and North Carolina, have various programs that identify the gifted and talented, and they carefully channelize their skills by placing them in specialized schools of technology and science. Among others, these offer the brighter students challenges commensurate with their higher abilities, and in the least, ensure that they do not regress to the average – mass mediocrity, if one may be more blunt.

## Teachers and Role Models

In the best institutions, the senior and most inspiring research professors do lecture to even undergraduate freshmen. They make room for undergraduate interns in their research programs. Many have cited the inspiration they drew from such professors as propellers of their own scientific and technological success. A close to home example for me is my own daughter who had the privilege of interning as an undergraduate with Dr. Doris Taylor, a pioneer in cardiac cell technology, and is now playing a lead role in a bio-tech company [24].

Feynman [21] has articulated the importance of such engagements with the young, and has been one of the best contributors to education through his Lectures on Physics. Feynman's statement that his classroom performance as a lecturer has more than once helped him regain his own self-confidence after some repeated failures in research attempts is particularly noteworthy. We need to honestly accept the fact that many who shun classroom engagements with students are not even the Feynmans of science. An educator needs not only to stand tall, but when needed, should willingly stoop to lift up.

When I was a student at Purdue, we used to have a research seminar each Friday afternoon in the auditorium addressed by famous mathematicians, statisticians, and computer scientists. The speakers were often the venerable who's who of science. The afternoon always ended with a social session that gave Ph.D. students an opportunity to freely interact with the visitor. Thesis professors eagerly encouraged their students to talk to these guests, informally, about their research. Such interactions strengthened my modest ability to separate a real contribution from a mere paper. Every time I see an educational institution in India excluding students from such experiences or students not taking the opportunities they get to interact, I feel sorry for the long-term opportunities being missed.

## Active Recruiting

Catching them young is key in many ways. In the developed world, many university research professors do take an active part in recruiting students. They even visit high schools and give lectures to get the young excited about science and technology, thereby, investing their time for the very noble cause of harvesting the best of innovation's seed corn. University and industry research organizations also arrange tours and in-house seminars for groups of students to give them a peek at the various types of research that takes place and the career options that they could potentially pursue.

## Education, the Key to Progress

For everything good in life and society, education reform is _the_ key to progress. It is much easier to build a new generation marching in the right direction than to try to set right an old one stuck in its own ways. The molding of the future happens in educational institutions, and educators are the true architects of that future. Despite the crucial role they play in molding young minds, it is unfortunate that teachers are often the least paid and rewarded (and held least accountable too) in most societies, including those that are capitalistic and highly developed!

## 2.5 The Government and the Bureaucracy

Technology innovation thrives in an environment supported by a government that is business friendly. From monetary and fiscal controls to various regulations and permits, the government impacts many parameters that affect the business climate of a country. If the actions of the government are not evaluated *a priori* and strategized to increase and support technology innovation, it becomes hard for innovation to thrive. Good governance at all levels is an essential lubricant to drive into high gear the commercial machinery including new business formation, and to prevent the process from getting impeded by inertia or friction. Indeed, developed societies with high levels of innovation thrive because their governments operate as service providers and not as masters, as enablers and not as controllers, and as facilitators and not as obstructers. Governments have to delicately balance the needs of business and society. Excessive controls can impede business and innovation. Excessive tilt towards business can render a government "of a people, by a people, and for a people" into one of a government unduly manipulated by business. Such transformations cannot lead to sustainable simultaneous economic progress and widespread improvement in the quality of life. Business is an important means to the worthy end of public good. The crony capitalism characterizing the US is an exercise in confusing the means for the ends, while the opposite, marking India, incurs the abdication of a set of important and crucial means to worthy ends.

## Law and Order

Whoever would believe that unconnected law and order related situations could impact innovation success? The intelligentsia in India does not connect well the divisive hysteria in the country with technological issues, and treat these as belonging to two distinct worlds, providing at worst only one more example of India's mind-boggling dichotomy. They tend to stay away burying their heads in the sand of academia or armchair politics. But as I write these very lines on November 26, 2015, entire India appears to have thrown itself in the middle of a controversy arising from the statement of a popular movie actor on perceived insecurity of certain minorities. Among others, it appears to have resulted in significant collateral damage to the success of at least one Indian e-commerce enterprise, for committing the innocuous sin of featuring that actor previously in its advertisements. The scientists and technologists of India have an obligation to comment at least as a group on this aspect and attempt to educate the public (no matter how uneducable they may think they are). Scientists and technologists must make a collective appeal to the government and to the opposition for a saner society. I, vicariously, cannot help seeing a lesson in this for technology companies too. They should celebrate more the achievements of scientists and technologists even in their advertisements instead of celebrities who are totally disconnected from technology.

## Red Tape

Time is the essence of entrepreneurial success. Particularly in technology and science, it is not enough to discover something important, but it is equally important to be 'first to market' and to establish market share quickly before competition heats up. A bureaucracy marked by one of red tape, power mongering, or corruption is a death trap for ongoing entrepreneurial efforts. It can choke the very stream and flow of entrepreneurialism by driving potential inventors and entrepreneurs to safer havens of paid employment. That may entail possible under-utilization, from a societal perspective at least, of rarely available talent. Worse still, this talent and its fruits may migrate to greener pastures in other parts of the world.

## Patents and Law

An inventor needs safeguards from infringements on intellectual property rights. It is not only important that trademark and patent agencies exist, but also that they inspire and keep the trust of confidentiality and function expeditiously. In some parts of the developing world, one does find significant resistance to even filing a patent application for fear that the material may prematurely leak out of government agencies. In some, the sheer burden arising from the lack of modern technology that could help with fast, efficient searches and easy electronic filing dissuades many inventors from applying for a patent.

Similarly, in places where copycatting is rampant, there has to be support both from law enforcement and the legal system. The adverse impact of corruption in law enforcement agencies and severe delays in court actions can tear the fabric of technology innovation in a country. These also dissuade foreign players in bringing to the country useful technology, some of which may themselves be useful drivers of greater domestic innovation and progress. Good governance in the IP area is a delicate act between actively encouraging new innovation and effectively controlling patent trolls and usurious profit mongers (both foreign and domestic).

Even if ready recourse is available through the courts, the cost of litigation can restrict the ability of smaller players giving an unfair advantage to large corporations. In the U.S., this has, to some extent, been remedied by third party financing of lawsuits, with hedge funds and other players willing to assume the financial risk (See [25]). It is essential for the society to evolve an intellectual property jurisprudence system that is equally accessible by all parties.

## Government Investment in R&D

For innovation in the very high-tech areas such as defense and space technologies, government investment with stability is absolutely necessary. As noted in Dr. Kalam's book [1], partnership with and outsourcing to industry pay large dividends by way of drawing a bigger circle of talent and speed-up of the entire process. India has certainly done an excellent job of this as one can see even from the open publications available in the market that chronicle India's nuclear and space programs. Unfortunately, it has also been my experience that these large contracts with assured payoffs combined with the easy money from body-shopping have also made it harder to get large Indian enterprises interested in taking even modest risks related to new innovations and ideas for the more competitive global markets.

In developed societies, the government is usually an unparalleled and easily accessible repository of information and guidance for the aspiring entrepreneur. They provide easy access to technical and other data, as well as patent and copyright related information. In addition, their governments invest in the creation and maintenance of reliable databases, electronic search and access, and networking to regional educational and research centers. Even the intelligence agencies of these governments have wings that serve the information needs of industry and research.

The government, in a typical developed nation, plays a significant role in setting standards. It is done not only to improve the quality of the nation's business output, but also to enlarge foreign markets through the sorting out of technology inter-operability issues. Sometimes a country might use its own specific standards, like ANSI, to protect the domestic market from foreign competition. This particular use of standards is a clever way of avoiding the imposition of laws limiting imports or imposing import tariffs that may violate various international agreements.

Embedded in the standards institutions are fine teams of researchers who work closely with industry partners and academia. These standards bodies also fund much academic research. Most importantly, these work as partners of domestic industry and not as an overbearing bureaucracy enjoying a sinecure. Indeed, developed economies with high levels of innovation are marked by significant collaborations among government, academia and industry that foster a consortium approach.

Despite making foul cries about government subsidies to business and championing anti-dumping when dumping is by others, developed nations do subsidize their agriculture and businesses in many ways through their research laboratories, defense related research contracts, and tax subsidies. The term 'research' is used in the corporate tax context in such a wide sense that much industry development efforts get subsidized by the taxpayer by companies renaming them 'research.' Whether governments are not wise enough or just wink is anyone's guess. Finally, a significant amount of research gets done by government agencies and gets handed over to industry on a silver platter giving the latter an added advantage in the global sphere.

## TABLE 2.2: Factors Affecting Innovation

| STIMULATORS | INHIBITORS |
| --- | --- |
| **Sociological:** | |
| Noble Discontent | Resigned acceptance |
| Desire for wealth | Wealth aversion |
| Willingness to change | Resistance to change |
| Risk tolerance | Risk aversion |
| Self reliance | Dependence on systems and structures |
| High Personal Accountability | Low Personal Accountability |
| High levels of probity | Corruption |
| Teamwork/extroversion | Excessive individuality/introversion |
| Rewards and recogntion | Lack of rewards and recognition |
| | |
| **Intellectual Property:** | |
| Awareness of value and IP rights | Lack of awareness |
| Streamlined dissemination | Lack of controls |
| Practical economic orientation | Sheer ego gratification |
| | |
| **Economy, Market & Industry:** | |
| High levels of competition | Lack of competition |
| Large consumer market | Lack of a large market |
| High disposable income | Low disposable income |
| Safety nets like insurance, social security | Inadequate safety nets |
| Industry R&D support | Lack of industry support for R&D |
| High Government investment for R&D | Low Government investment for RD |
| Efficient Distribution Networks | Lack of distribution channels |
| Hierarchical systems of management | Participative management |
| | |
| **Finance:** | |
| Low inflation | High inflation |
| Low cost of money for R&D | High cost of money for R&D |
| Venture capital availability | Venture capital unavailability |
| Private equity availability | Inadequate private equity |
| | |
| **Education:** | |
| Abundant high quality institutions | Lack of high quality institutions |
| Orientation towards merit | Low merit orientation |
| Multi-disciplinary structures | Silo like structures |
| Good theory / application mix | Bad theory / application mix |

**Table 2.2: Factors Affecting Innovation** - Contd.

| STIMULATORS | INHIBITORS |
| --- | --- |
| Innovation oriented | Without specific direction |
| Integration of eng. and business | No integration |
| Interactive instruction | Unidirectional instruction |
| Intellectual orientation | Hierarchical orientation |
| Adequate access to senior researchers | Inability to interact |
| Exploratory learning | Rote learning |
| Focus on problem solving | Focus on theory only |
| Entrepreneurship skill building | No entrepreneurship training |
| Streamlining of promising talent | No streamlining |
| Co-existence of hands-on work | No hands-on work |
| Faculty with industry experience | Lack of industry experience |
| Industry internships | Lack of industry support |
| Industry partnership | Lack of industry partnership |
| High level faculty accountability | Entitlement based systems |
| Encouraging reward structures | Lack of reward structures |
| | |
| **Government:** | |
| Service Provider | Ruler |
| Enabler | Controller |
| Helper | Obstructer |
| High integrity | Corruption |
| Business friendly | Business antagonistic |
| | |
| Good IP protector | Poor IP protector |
| Trustworthy IP agencies & staff | Dubious IP agencies & staff |
| Effective Law Enforcement | Corruption and Delays |
| | |
| Partnering with private sector | Stifling private sector |
| Providing business with subsidies | Providing little subsidy to business |
| Consortium with industry & academia | Disconnect with industry or academia |
| | |
| Investing for innovation | Inadequate investments |
| | |
| Repository of data & know-how | Inadequate as a repository |
| Providing support systems | Providing inadequate support |
| Standards bodies with business focus | Weak standards bodies |

## Notes and References

[1]   A.P.J. Abdul Kalam: *Wings of Fire, An Autobiography,* University Press, India, 1999.

[2]   Entrepreneurship and the U.S. Economy, http://www.bls.gov/bdm/ entrepreneurship/entrepreneurship.htm

[3]   James Grubman: *Strangers in Paradise, How Families Adapt to Wealth Across Generations,* ISBN-13: 978-06-15894355, Paperback, Family Wealth Consulting, www.jamesgrubman.com, 2013.

[Though written by a financial psychologist for the wealthy to ensure retention of that status by subsequent generations, my observation is that certain chapters of it pertaining to attitudes towards wealth have much relevance at the macro level as well. This is an interesting topic for research in itself.]

[4]   Adi Sankara, considered one of the foremost exponents of Hinduism was an ascetic who wrote much in praise of the path of renunciation. However, he took special pains to note in his commentary on the *Bhagavad Gita,* the sacred book of Hindus, "Twofold indeed are the *dharma* adhering to *vedas,* which form the very basis for the sustenance of the universe: those involving action (acquisition); and those involving cessation of action (renunciation)" and asserted that one of the goals of *Vedanta* is "providing (also) welfare in a worldly sense (*prāninām sākshāt abhyudaya nihshrēyasahētu:*)." See *Srimad Bhagavad Gita Bhasya of Sri Sankaracarya,* A.G. Krishna Warrier, Sri Ramakrishna Math, India. ISBN 978-81-7823-507-3.

[5]   Economic Moat, Morning Star, http://www.morningstar.com/Inv-Glossary/economic_moat.aspx

[The term 'Economic Moat' refers to the ability of a business to fend off competition over a long period. Note that one of the essential building blocks of that is intellectual property capital.]

[6]   Peter Temin & Louis Galambos: *The Fall of the Bell System,* Cambridge University Press, 1987.

[7]   John Gertner: *The Idea Factory: Bell Labs and the Great Age of American Innovation,* Penguin Press, 2012.

[8]   "Why Aren't There More Female Patent-Holders?" Freakonomics, March 7, 2012.
http://freakonomics.com/2012/03/07/why-arent-there-more-female-patent-holders/

[9]   J.Hunt, J.Garant, H. Herman &D.J. Munroe: "Why Don't Women Patent?" NBER Working Paper No. 17888, March 2012. http://www.nber.org/papers/w17888?ntw

[10]  Joe Fassler: "How Doctors Take Women's Pain Less Seriously," *The Atlantic*, Oct 15, 2015. http://www.theatlantic.com/health/archive/2015/10/emergency-room-wait-times-sexism/410515/

[11]  Maya Dusenbery: "Is Medicine's Gender Bias Killing Young Women?" *Pacific Standard*, Mar 23, 2015.

[12]  "What is 3D printing ?"  http://3dprinting.com/what-is-3d-printing/

[13]  Mark Boroush: "NSF releases new statistics on business innovation," http://www.nsf.gov/statistics/infbrief/nsf11300/

[14]  Geoffrey Moore: *Crossing the Chasm: Marketing and Selling High-Tech Products to Mainstream Customers,* Harper Collins, New York, 2002.

[The chasm referred to is the big gap between the early adoption of a new product by a small number of enthusiasts and the latter acquisition by the product of a large market share.  The inability to cross this chasm successfully poses one of the highest dangers to a firm bringing out a revolutionary new product.]

[15]  Gregg Fairbrothers & Tessa Winter : *From Idea to Success, The Dartmouth Entrepreneurial Network's Guide for Start-ups,* McGraw-Hill, 2011.

[16]  Andrew Romans: *The Entrepreneurial Bible to Venture Capital,* McGraw-Hill, 2013.

[17]  Robert Finkel: *The Masters of Private Equity and Venture Capital,* McGraw Hill, 2010.

[18]  David A. Vise: *The Google Story,* Bantam Dell, Random House Inc., New York, 2005.

[19]  Lucas Carlson: *Finding Success in Failure: True Confessions from 10 Years of Startup Mistakes,*  Craftsman Founder, 2015, ISBN 978-0-9960452-2-3.

[20]  John Lusk & Kyle Harrison: *The MouseDriver Chronicles: The True-Life Adventures of Two First Time Entrepreneurs,* Basic Books, 2002, ISBN-10: 0-7382-0801-9.

[21] Richard P. Feynman, *'Surely You're joking, Mr. Feynman!' Adventures of a Curious Character,* 1985, Bantam Books, New York. [See the section titled, "O Ameriano, Outr Vez!.."]

[22] George Otto Trevelyan, *The Life and Letters of Lord Macaulay,"* Longmans, Green & Co, London, 1876.
https://archive.org/details/lifelettersoflor01trevuoft

[23] "New Couse at Harvard Business School,"

http://www.boston.com/business/news/2013/09/30/new-course-harvard-business-school-product-management/DHriTMYDnvhs11776lXRVJ/story.html

[24] RoosterBio Inc., http://www.roosterbio.com/

[25] "Caterpillar hit with $73.6 million trade secrets verdict in U.S.," http://finance.yahoo.com/news/caterpillar-hit-73-6-million-171331289.html

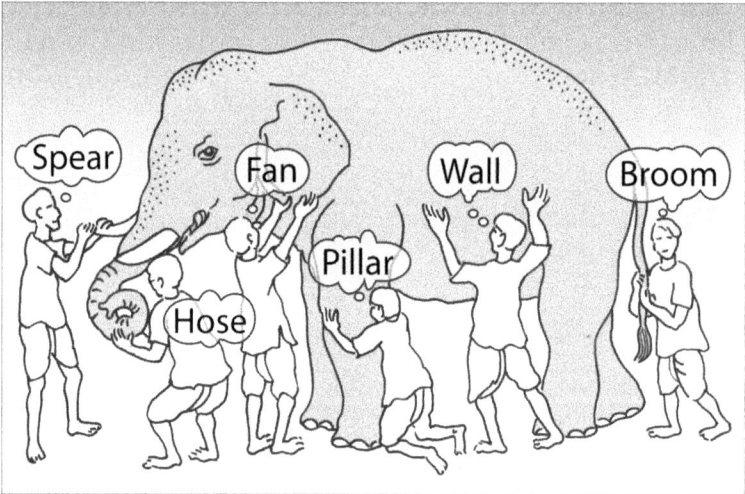

Figure 3.1 The blind men and an elephant

## Chapter 3

# India's Highly Disconnected Society

We consider, in this chapter, some important social factors discussed in Chapter 2, but in the specific context of India. Each one of the factors discussed here impacts Indian innovation, product development, and new business formation directly or indirectly. For example, corruption deters many from taking the necessary risks associated with entrepreneurship. Black money and conspicuous consumption, particularly of imported luxury items, divert much needed resources for capital formation. While societal factors are perhaps the hardest to control, and even though certain changes may occur only over a long haul, yet there are many aspects that can be addressed in the short run to yield tangible results. In any case, the longer the country waits to start addressing the issues earnestly, yet longer will it take to resolve them. The challenge will only get harder by problems getting worse over time. Hence we include these in our discussion.

The reader should, however, be forewarned that anyone describing India runs the risk of appearing like one of the proverbial blind men in Figure 3.1 who sets out to describe an elephant based on the feel of only one of its parts. India's diversity is mind boggling to an extent that the well-known consumer researcher Bijapurkar [1] called (consumer) India "totally schizophrenic."

> "And it's not just the twenty-three languages, the geographic and climatic diversity, the different religions living together, and the many shades of rich and poor people that exist, or coexist in this vast continental country. India has twenty-eight states, and there are wide income and social development disparities among them." - Bijapurkar [1]

India is home to a number of highly westernized, jeans wearing, disco-dancing, latest iPhone or Android toting, rich youngsters, and simultaneously to a large number of monetarily impoverished people who are, sometimes, even forced to miss a meal. (For a comparison, note that the price of about Rs. 50,000 for the latest iPhone or a comparable Android device can feed more than 1000 adults for a day in some rural areas of India.) There is the high likelihood of seeing right next to a posh, gated, high-rise residential complex with tennis courts, swimming pools, and other amenities, a slum where barely clothed children play in absolute squalor. A high-tech research institution of international fame may have to be reached only via a roadway on which bullock carts vie for right of way with a latest model BMW. Changing this requires many changes in society and cannot be accomplished if business remains as usual. One of the drivers of such changes could be innovation. Sengupta [2] gives several recent examples.

However, societies, particularly democratic ones, do not lend themselves to forced social engineering, and the state cannot mandate changes that bring quick results. Attitudes and behavior entrenched over decades do take a long time to change. Frustrating though that may be, it is certainly good to know the lay of the land, and, where possible, to try to influence at least future generations and those sections of society that have a higher propensity to change and become major drivers of progress.

## Resigned Acceptance

India is the very antithesis to the 'noble discontent' we noted in Chapter 2 as one of the two major drivers of innovation. The attitude of the majority is one of resigned acceptance marked by refrains like, "We are like that only" (chosen as the title of her book by Bijapurkar [1]) or "What can we do? Adjust!" when even correctible issues are brought up for discussion.    Although one does see occasional and sporadic riots and upheavals, those are mostly local and serve, at best, as short fused cathartic steam letting by the young or the poorer sections of society. Unfortunately, these rarely bring lasting changes as those in power quickly douse them by force or through some handouts. The vast majority of the educated middle class, who could be the engine of progress and change, are, however, apathetic, uninvolved, and totally resigned. They accept everything from poor quality of products and services to blatant and open corruption that often borders on extortion. The rich people do not protest or put up a fight as they can buy their way most of the time.

Some have tried to explain away this willingness to put up with the current state of affairs based on the so-called fatalism of Hindus who form the majority in India. Hinduism preaches one to accept everything that comes as a gift of God (*iswara prasad*). It also explains away one's misfortune as a consequence of the actions (*karma*) of many previous births that have to be endured in the current one. But one should discount any argument based on Hindu religion for many valid reasons. First of all, most Hindus understand perfectly that these exhortations of the religion are only to develop a healthy dispassion towards the results of one's actions and to maintain internal sanity in an imperfect world.  They understand that it is not a call for inaction. Secondly, the hold of religion, like in most parts of the world, is diminishing significantly in India too, especially among the educated young and the urban population.  Finally, the response of resignation and apathy in India is in no way unique to the Hindus.

## Corruption and the Culture of Graft

The prevailing passive acceptance is rooted in a sense of real impotence experienced by the educated non-rich in an environment characterized by a high degree of nexus between politics and hooliganism. An increasing number of states in India are marked by blatant corruption even at the highest levels of government. Suppression of opposition and complaints, using both police and extra-governmental force, is all too common. The level of corruption of politicians and high-ranking officials in some states of India may make the robber barons of the 'Gilded Age' of America [3] look like saints!

We have already noted in Chapter 2 the need for a high level of probity in the society across the board for innovation to flourish and for new businesses to be formed. Corruption is a universal phenomenon in the world, but developed societies have learned to keep it in check from impeding their day-to-day life or businesses. Contrary to that, the lack of probity has become so widespread in India that it affects all activities and all people in all sections of society. It drives the nation's highly talented either out of the country or into sanctuaries of paid employment (often with foreign multinationals), both resulting in what is, undoubtedly, an incredible amount of an opportunity cost to the nation. Among those impacted are wealth creation through indigenous innovation, product generation, and new business formation. That necessitates somewhat of a longer discussion of corruption.

In India, those in power perpetuate their existence through control of large vote banks of the uninformed poor by handing them various gifts funded by 'donations' extorted from businesses. Fanning various types of class animosities is another successful tactic. The educated classes are painted as upper and privileged castes from whom the masses have to be protected. They encourage their supporters to openly flout laws and property rights by condoning acts such as illegal encroachments. Courts are ineffective too, and instances of cases (especially when the accused is rich and powerful) remaining unresolved for decades are commonplace. Most enforcers of law are themselves corrupt, and the few honest ones end up paying a high price [4], [5].

Nothing gets done without someone's palm being greased. What distinguishes India is not corruption or its ubiquitous presence, but that even to do what is well within the law and to get even what is legitimately one's due, one is often forced to pay a bribe to someone. Even simple things like obtaining a marriage certificate or registering a property cannot be accomplished in most places without giving a bribe. Thus, even the most unwilling are forced to participate in the corruption, at least as bribe givers for essential needs. To say that in today's India it is almost impossible for one to be honest would not be an exaggeration at all.

The people of the educated middle class should be empowered to value honesty by providing them proper legal protections. They should be brought back into politics and government at all levels. Proper avenues should be set up for airing and redressing legitimate complaints in a timely manner. Without these, the situation will not change. It will significantly impede the progress of the nation and condemn it to remain, for many more generations to come, among nations of the world with the lowest quality of life.

The educated middle class is not to be absolved entirely of the moral decadence either. Even as they decry corruption loudly, many of its members benefit from corruption in diverse ways. They have come to value wealth a lot more than how it is acquired, and that is reflected in the way they conduct important personal matters including those relating to arranged marriages and forming other alliances.

Although most bribes are extorted by explicit or implicit threats and are given unwillingly, much bribery also occurs for gaining special privileges or to evade the law or taxes. There is a large parallel economy in India of unreported wealth that is sustained by official corruption. But, with only a small fraction of the population having official power to extract bribes or the economic means to give bribes, the benefits of this system accrue to a small number at the expense of the vast many. True uplifting of the masses simply cannot occur without getting rid of corruption. Such uplifting is, however, absolutely essential in India to provide a large market for domestic products and thereby to encourage domestic innovation. Furthermore, continuation of the present situation is a prescription for mass eruptions at uncontrollable levels threatening the very nation state.

The greatest need of the country is to educate the masses about true democracy in a way they can keep self-serving corrupt politicians out of government at all levels. At present, in many states, the available choices of parties are equally bad. Even if the party in power changes, business goes on as usual. Those that are newly elected never bring to justice opponents who were corrupt and got removed from office by the electorate. This could be because they are afraid that their own turn may come later. Obviously, technology offers the means to effect major reforms through significantly improved methods of monitoring and policing, as well as by public education, but the decision to deploy those means rests with the very corrupt who have a major vested interest in maintaining the status quo.

There is a glimmer of hope, however, due to the advent of social networks, webcams, portable recorders and the like. But, presently, most of that technology is used for exchanging trivia and mostly by the non-intelligentsia. As for the caring intelligentsia, newspapers, and news media, they live in an overall atmosphere of fear and intimidation that prevents them from being willing to use the technologies to drive reforms.

It was a grave mistake on the part of India's Constitution framers to have left law and order almost entirely in the hands of the state governments. This has allowed many local despots to monopolize power and amass personal wealth even more than the despotic *rajas* (petty kings) and *zamindars* (landlords) of the past, at the expense of the nation. They have also bred a whole class of civil servants whose income from bribes far exceeds their government salaries, and whose qualifications and contributions are marginal. Nepotism and bribes play a major role in government appointments and also in admissions to schools and colleges. Even the private sector is forced to oblige in such matters bringing an overall deprecation of institutions including the prestigious ones of a past era.

Existing systems of enforcement have shown an eminent inability to deal with the situation and are themselves subject to the same ills in the larger society they are supposed to curb. The central government needs to enact a set of special laws that hold across the nation to curb corruption, appoint special agencies free from political intervention to enforce those laws, and special courts to punish and remove offenders from office expeditiously. The same holds for acts of bias based on gender, religion, caste, or state of origin. Equal justice to all sections of society alone can solidify the nation state and restore faith in law and justice and in the various systems set up to enforce them. The first place to start is, of course, the Central Government, where bureaucracy is rampant. Steps should be taken to create an environment where corruption receives zero tolerance, and abusers of power are punished swiftly and severely.

## Subordination of Merit

India's mediocrity is, to a large extent, self-made and owes to the subordination of merit in all aspects of life. Every job in government and seat in an educational institution is viewed as an opportunity for a bribe by both politicians and civil servants. Nepotism also plays a big role in cornering them. The tentacles of the corrupt politicians and officialdom extend also to private industry, which is forced to take in a quota of less qualified employees and to provide special preference to relatives of those in power and "sons of the soil" even if they are not the best qualified.

Added to all this is the reservation of seats based on caste and religion with no regard to maintaining high standards of merit even in the once elite institutions. Although some reservations were enshrined in the Constitution of India with the good intention of setting right the injustices of the past, particularly to the 'untouchables,' they were to have been removed gradually. Instead, reservations have been widened indiscriminately (primarily to create vote banks), and merit thresholds for acceptance have been lowered for some groups to fill assigned quotas. It has become clear that other than enriching enormously a small number of corrupt politicians and members of their immediate circle, this has done little to uplift the masses. In the meanwhile, there is a great influx of less qualified persons across the board in government and in many professions affecting highly the very people these schemes were intended to help.

A strong case exists against the current system of reservations based on caste, setting up quotas for subgroups, and the lowering of merit thresholds to meet the resulting quota. (a) Although they may benefit some individuals, collectively the nation loses through the mediocrity it breeds. (b) The presence of mediocrity affects morale and performance across the board creating a regression to systems that are highly suboptimal. (c) The policies do not help the economically disadvantaged people. (d) They cast an aspersion on the meritorious members of the 'backward' groups as having benefitted by a dole. (e) The overall quality of products and services within the country degrades, and that worsens the situation for all. In fact, it affects the poor more through scarcity of quality products and services and the resulting increase in prices for them. There may have been a case for reservations and quotas earlier from the perspective of representation and participation in decision-making bodies. But now, given the levels of infusion of various groups into the mainstream, a time has indeed come when reservations based on caste and the quota system need to get phased out, at least over the period of a few years, by reducing gradually the number of seats reserved and by increasing gradually the thresholds of acceptance. Eventually, there should be one threshold that applies to all.

It is my firm belief that, in future, economic status should be the primary basis on which the very term 'disadvantaged' is defined. The goal of preferential schemes should be to remedy the causes that contribute to an uneven playing field so that the previously disadvantaged can compete as effectively as the others. For example, children in poor slums may come relatively less prepared for schooling because they do not have access to various learning toys or to an environment that allows pre-school learning at home. The right way to remedy this is to set up centers in poor neighborhoods so that the children in these areas can get from them the same pre-school learning as their more fortunate counterparts. In addition, such centers could offer counseling services to the parents so that they can create a home environment that will prepare their children for school. The present system of entitlements based on improper criteria and without regard to merit or personal accountability only degrades the playing field for all. It is not geared to accelerating the nation's economic progress to a level it can support a climate of high innovation.

## Hierarchical Structures

Indian society is highly hierarchical. It is as though everyone works *under* someone, and not *with* others or *for* themselves, their family, or the nation. Within many families, there is a strongly pronounced hierarchy based on age, with the younger members expected to accept and obey instructions without question or inquiry. There, and in most schools and colleges, even legitimate inquiries are often treated as acts of insubordination. These practices have a serious effect on independent thinking that is essential for innovation. Few work environments allow for any upward communication from workers. That leads to a loss of valuable information from those who have a first hand knowledge on the job floor on how things could be improved. An environment fostering a spirit of inquiry and openness with a right balance of structure and authority is absolutely essential for India to become more innovative. In the society and at all its levels, recognition must occur that respect does not equate with implicit, unquestioned obedience or servility.

The hold of hierarchy extends into academic circles as well. No conference can take place in India without a long opening session paying obeisance to a whole host of local chieftains who must also be allowed to speak. Several plenary slots get hogged by senior professors even when they have no recent research of their own to report. Younger colleagues get pushed to end of the day or end of the week sessions even if they have worthwhile papers. It is a travesty to deny deserved visibility to younger colleagues who have done good research. Some years back, as a co-chairman of the International Teletraffic Congress (ITC) held in Washington, DC, I successfully managed to break that sort of a practice at the ITC with a new format where most plenary slots were allotted to the best among the submitted research papers and the rest to a few specially invited speakers to speak on very new technologies. Many seniors who previously enjoyed plenary slots as entitlements were invited to serve only as session chairs. That change was uniformly well received and became the format of future ITCs. When I proposed a similar format for an international Indian professional organization in the USA, the result was, however, just the opposite. I got sidelined forever, despite being a life member. Luckily, I was not at a stage of my career where these things mattered. It is somewhat sad that hierarchy is so entrenched in Indian thinking that many carry that baggage wherever they go.

## Dignity of Labor

Among the many changes that India has to make in order for it to be successful is its willingness to demonstrate greater appreciation for the dignity of labor. People deserve basic courtesy and kindness irrespective of the work they do to earn a living. Unfortunately, that is not the case in India. As an example, consider this. Most Indian languages have two forms of address comparable to the "you" and "thou" of English, and the "*vous*" and "*tu*" of French. The "*aap*" and "*tum/ tu*" of Hindi are Indian examples. The latter form is used routinely in India by those higher up while addressing those at the lower end, with little regard even to the age of the person with whom they are talking. The poor worker rarely hears words like "Please" or "Thank You," but, instead, is subjected to curses and scolding that no educated person with some means would tolerate.

At a mass level, the lack of appreciation for the dignity of labor has resulted in everyone wanting to get degrees and white-collar jobs in the hope these will spare them these indignities, although in most instances they do not. The flip consequence of this is hyper-competition for college admissions and white-collar jobs fanning corruption and merit erosion, a dearth of qualified labor of all types, and even innovation to improve tasks and occupations at the lower end of the spectrum.

The worst example of the ills of social hierarchy in India is the area of manual scavenging [6] that, despite all attempts of the government, seems to still persist. How unfortunate that Indian society has not adapted or innovated technology and tools that extricate the scavengers from their despicable plight!

## Resistance to Change

While the top few percentages of the population with much higher income than the rest may be chasing the latest products and fads, the bulk of the population is very resistant to new technology and is quite content to carry out their tasks with whatever inefficient methods and tools they are accustomed to. Much of the lure of India for foreign multinationals comes primarily from the top 10% or less of its population who can afford their products, and the sheer number that even 10% of India translates into. The consumption pattern of these fortunate Indians is skewed heavily towards snob items like super luxury vehicles, expensive handbags, watches, and cosmetics (that should all, of course, bear a recognized foreign brand label). The economic liberalization of India has certainly been a boon to foreign makers of these products, but has done little to accelerate India's own indigenous rate of innovation or product creation. Even when some innovations like selling items like shampoo in small sachets have been introduced to the "bottom of the pyramid" in India, the driver has been mainly profits of some large corporations.

Unfortunately, the wealthy and the upper middle class that spend their monies on luxury and snob items often fail to recognize that it would be more sensible to spend their monetary resources on those products that could drive higher productivity and even innovation and business creation in the country. Let me cite a few examples. A typical upper middle class family in India would struggle everyday berating maids who do not show up regularly or do not perform their jobs well, rather than buy a vacuum cleaner they or the maid could use. Similarly, people would be reluctant to switch to a good serrated knife with an ergonomically designed handle to cut vegetables, but would be more in favor of using a blunt flat knife just because they are used to the latter. Even in parts of the country where monsoon rains cause severe problems for clothes drying, the possibility of seeing a clothes drier, even in households that can afford them, is unlikely. Many contractors will not invest in things like high quality tools, and certainly not power tools, for fear that they may get stolen or be broken by misuse. Poor accountability, inability to get quick legal redress, the poor reliability and high cost of electrical power are all certainly issues that get in the way for many, but the more important causes are the overall willingness to put up with substandard work and an argument that labor is quite cheap anyway. India needs to break this logjam to move forward into being developed.

Drèze and Sen [7] correctly argue that much of the economic liberalization by the government has helped only to satisfy the rich and the upper middle classes, and that a large section of the country is left out with no participation in the new economy. An inclusive approach could actually help to drive the economy to high gear. The lower echelons should also be empowered equally to become consumers and producers themselves. Economic liberalization has to be liberalization of all sections of society and should be accompanied by educating rich consumers of the larger picture beyond conspicuous consumption that only impoverishes the nation and drains its foreign exchange reserves.

## Risk Aversion and Lack of Safety Nets

The bulk of the Indian population is risk averse. Less than 1.5% of the Indian households invests in securities. Their investment in stocks accounts for a mere 2% of all household savings. In contrast, 10% and 18% of households in China and the US invest in securities; the long run average of 45% of US household savings goes into securities [8].

A significant part of India's household savings lies in the form of gold [9], [10]. The total amount of gold held by Indian households is of the order of 20,000 tons and valued around 950 billion US dollars. An attempt by the government to turn that into productive use has met with little enthusiasm [11]. This and the amount of black money that does not enter the economy in productive ways impedes capital formation in the country significantly and, in turn, affects innovation and new product development.

There are many reasons for wealth getting hoarded as gold, but the most important ones are the high volatility of the Indian stock market, some past occurrences of scams, and a general perception all around (and an irrefutable fact) that India's stock markets are highly manipulated and subject to the whims of foreign investors and their actions. Although markets are regulated and strict laws do exist on the books, enforcement is poor. The overall level of corruption in government and various enforcement agencies dissuades involvement in the market except by a small few. The court system with its unbelievable delays does not help either. Some foreign financial houses have also cited the substandard ability of Indian stock pickers to make good investment choices for their clients. Unlike in countries like the US, the common man has little access to detailed information on or unbiased analyses and comparisons of companies. For instance, there are no subscription services, comparable to a MorningStar or Zacks investment research, which are credible and affordable to the investing public.

Indians lack safety nets in the form of Social Security and unemployment payments that are common in developed nations. A serious illness or death of a wage earner can throw a middle class family overnight into severe poverty and dire need. That forces households to save in secure investments. That, once again, limits the capital available for businesses.

## The Curse of Black Money

A very large amount of India's wealth, estimated to be "between $400 billion to more than $1 trillion" [12], is totally black and unaccounted for. It is earned, hoarded, and spent in many ways under the radar. Some of it gets spent in lavish weddings and in other ways that are beyond the eyes of the taxman [13] and form yet another opportunity for bribes. A sizeable chunk is also stashed in countries like Switzerland whose banks shamelessly serve as well advertised brothels for the financial prostitution this entails. The activities of such countries and their banks need to be put on par with terrorism, for they indeed are a form of economic terrorism. Severe sanctions should be imposed by the international community both on these financial institutions, and the governments that wink at their insidious practices. Without a concerted effort by the entire international community, the only ones who are able to benefit from even major leaks like the infamous WikiLeaks are big economic powers like the US. Other countries have little leverage internationally [14]. Internally, they also lack the will to stop this menace because many of those parking black money in such places are politicians and the super rich who are "too powerful to jail." Like in many other matters, countries like the US have acted only to the extent of securing their own interests without honestly addressing this global menace.

India is seen increasingly as a potential savior of many rich exporting nations whose populations are now ageing and economies beginning to totter. They vie with each other for India's market. One would assume that India will use that as a leverage to get greater cooperation from them in matters of great importance and lock them out of trade, at least of discriminatory items, if they do not cooperate. But in India, politicians and policy makers are for sale especially to foreign buyers, and there is no will to exert the influence that the country really can exert.

India needs substantial investment funds to ignite its businesses and productivity, and to achieve and maintain high growth rates. But if India's growth is to occur without surrendering much ownership of businesses and assets, including intellectual capital, to foreign nationals (including NRIs), it is absolutely essential for India to address the above issues and make noticeable progress quickly. The present situation locks up significant resources of the country in passive, non-income producing assets. It robs the exchequer of taxes, and inhibits the internal investment climate severely by adversely affecting business activity, new business formation, and certainly capital needed for domestic innovation and product development.

## Idealism Tempered by Pragmatism

Over its entire history, the ethos of India has been influenced highly by its commitment to non-violence and non-aggression. That ideal has resulted in a voluntary abandonment of some strategic options including a first use of weapons of mass destruction. The result of that, however, has been that India is forced to spend an enormous fraction of its GDP for defense outlays at the expense of national development. The repeated indiscriminate arms sales by the US and others to certain neighbors who have shown an inability to be good citizens of the world force India to maintain a large conventional force and to import expensive defense equipment of various types. From the perspective of enabling the availability of greater monetary resources for overall national development and innovation, India needs to revisit some of its idealistic policies and temper its idealism with a proper dose of pragmatism. That alone would enable it to get out of an arms race and yet defend itself fully. This is especially feasible today given that India has not only weaponized key technologies but has also developed credible delivery systems, both strategic and tactical.

## Effects of Globalization

The buzzword, today, is globalization, and it is touted and pushed heavily by predominantly exporting nations, the United States being their self-appointed leader. Both within countries and among them, the effects of globalization have been mixed. The insightful book [15] of the Nobel Laureate Joseph Stiglitz details how globalization has "the potential for doing a lot of good" but "has not been pushed carefully or fairly." India has also had a mixed bag of experience with globalization.

The two largest democracies of the world, the US and India, owe, to some significant extent, their independence to the realization of their leaders and thinkers that they need to break the vicious cycle empowering the oppressor by buying more of their goods and services. The US had its 'Boston Tea Party,' and India its '*Swadesh* Movement.' The US is now beginning to re-learn that lesson with respect to China and some Middle East countries, while India remains entirely oblivious to that need.

Prior to independence, India suffered many strictures from Britain, its colonial master, which actively impeded India's industrial growth. When India became independent, its early leaders like Nehru and economists like Mahalanobis were highly cognizant of the need to create an environment in which Indian industry and businesses could develop. The government was protectionist in many ways, and rightly so. They also opted for a planned economy and policies driven by socialistic goals. Most importantly, Nehru tried to maintain a stance of 'non-alignment' with the opposing super powers without allowing the country to get dragged into their proxy wars. (Look at poor Pakistan that did not have that wisdom). They were willing to settle for a slow and steady pace of inclusive development, instead of an unsustainable flash-in-the-pan type of evanescent prosperity of a few and the ownership of most of the country's assets by foreigners once again.

There is much debate today on whether and to what extent the protectionist socialist policies and 'non-alignment' helped. Later economists, including some Indians like Raj Krishna who coined the term 'Hindu Rate of Growth' [16], may deride them to their hearts' content, but it is undeniable that India could not have built a strong infrastructure, a sizeable educated class of workers, and the foundations for a strong economy without those policies in its early decades of independence. It is also possible that the country would have erupted in a civil war if it had pursued a capitalistic structure at the risk of its poor not benefiting from independence and being left to remain as they were in the British Raj. It is important to note that there is no credible counterexample of a country in the world that one can cite to discredit the choices of early India if the criteria are based on sustained and continuous growth, freedom of the people, social justice, domestic ownership, and ability to gain respect and exert influence in the international arena.

Although it may be heretic in some circles to say so, nationalization of banks in India brought more money to small businesses and farmers. That, and the empowering of the farmer by cutting out the middlemen through nationalization of grain trade, helped the green revolution. These spread economic progress beyond urban areas and to a larger set of people. Indeed, many benefits did accrue from the socialist policies of the government.

Despite the steady progress in many areas, the speed of overall progress and growth was not just slow but very slow. Corruption and inefficiencies increased over time and got in the way. This could be attributed more to the fact that the idealism of the leaders did not percolate down to their followers and implementers, and less to any inherent defects in the approach itself.

Certainly, the policies of Nehru and his cohort were not sustainable in the long run, for inherent in their emphasis of public ownership and control was a fertile ground for corruption and the operation of the familiar dictum, "Everybody's business is nobody's." Also, the world order has changed significantly from those times. The cold war became a thing of the past leaving only one super power. So, various changes became necessary, and it is good that a fair amount of liberalization has indeed taken place. The big question is whether India is going too far without honest introspection and an appropriate level of importance attached to falling into the trap of crony capitalism [17], [18] and a new form of colonialism. Also, are we evaluating the effects of liberalization taking into account all segments of Indian society?

One may argue that the so-called economic liberalization in India has spurred much economic activity and created a large middle class. But the repeated de-valuations of the Indian currency have made India poorer in many ways. Our exports have not kept pace with the increasing levels of imports. That has worsened India's balance of payments, and if one takes out inward remittances by NRIs and the earnings of the IT sector, the disparity would become even more glaring. When economic measures are taken not in PPP (purchasing parity power) terms but in real exchange terms, India has not improved much at all. The typical Indian worker works much longer hours and enjoys only a fraction of the quality of life enjoyed by his peer in the developed world. A consequence of the devaluations is that the sweat and blood of Indian labor enrich many foreign nations by providing cheap services and products making it even harder for India to catch up. Yet, India continues to face tremendous obstacles for its exports from the very same countries that view it only as a market for their goods. Nonetheless, India's own leaders seem to emphasize a lot more the labor and the market India can offer to the world without an equal emphasis on India's indigenous growth and production [19]. A high degree of emphasis should be placed also on technology transfer to India and ownership by Indians in India. This is a matter of serious concern for those of us who have seen and understand the plight of the peoples of various banana republics of the world. Given India's intellectual assets, India deserves to be a lot more than becoming a provider of cheap labor to the developed nations.

## Intellectual Assets

Amidst all the doom and gloom of India, what really gives tremendous hope for the country is the large pool of its highly talented and educated scientists and engineers. An example of what the nation can achieve by empowering them is already seen in the area of Software and Information Technology, which has contributed enormously to the creation of a large middle class in India and has put the country on a path of ascendency. Dr. V. Rajaraman [20], a pioneer in computer science and winner of a Padma Bhushan award by the Government of India for his own contributions to it, has given an excellent and highly readable account of the history of that development as well as of computing in India in general.

The history of computing in India is worth recalling at least briefly here, for it has many lessons to offer. Much development occurred under the auspices of the Department of Electronics, formed in 1970 by the Government of India and led by M.G.K. Menon, a physicist and former Director of TIFR. Early support from it (including sponsored study abroad) helped a number of Indians acquire a high degree of knowledge on computer technologies. In cooperation with TIFR and IIT, Kanpur, the computer division of India's public sector enterprise Electronics Corporation of India Ltd (ECIL) developed various models of mini-computers, an operating system, an assembler, and a FORTRAN compiler. Although these did not succeed commercially in a big way, nevertheless they helped to train over a thousand engineers in designing systems and developing systems software. Many other efforts of the government such as the establishment of the National Center for Software Development and Computing Techniques (NCSDCT), National Informatics Center (NIC), the Army Radio Engineering Network (AREN), and Air Defence Ground Environment system Development (ADGES), as also the establishment of Regional Computing Centers (RCCs) with main frames in university campuses, created a very large pool of people with significant expertise in software design and systems engineering.

A great opportunity for indigenous development opened with the exit in 1978 of IBM who would not agree to the restrictions imposed by India on Indian co-ownership of its Indian operations. Simultaneously, for the first time a course on Computer Sciences at the bachelor's level was started at IIT, Kanpur by Professor Rajaraman, whose immediate popularity was to result in the proliferation of such courses all over India's institutions of higher education. In parallel, most typewriting institutes, which saw impending obsolescence of typewriting and stenography, morphed into computer classes teaching programming, and yet another quiet revolution occurred in India unaided by the government. Then the Y2K fear (the fear of the dawn of the year 2000) hit the developed world, and that needed the recruitment of a large number of people to wade through miles of computer programs to detect if the use of only the last two digits for designating the year would create serious problems and bring much of the computerized world to a grinding halt. Much of the legacy software was in COBOL, which the new generation of programmers in developed countries hardly knew. But, many institutes and colleges in India taught COBOL. They also taught an antiquated version of the dreaded JCL, which few continued to learn in the West. In addition, Indian labor was relatively much cheaper for the developed world.

In a rare instance of technology backwardness becoming a blessing, what started as the capture of an opportunity in software consultancy in some small measure by some companies like Tata Consultancy Services (TCS) ballooned into a big enterprise with many new companies in India being formed, and the big industry of exporting programmers grew out of it. That growth was accelerated even more by the government through the provision of infrastructure like Software Technology Parks with satellite communications and tax concessions of various sorts.

The blockage by the US government of the purchase of a Cray supercomputer led to the development of super computer technology in India and the creation of the Center for Development of Advanced

Computing (CDAC), which built a super computer called PARAM PADMA that got ranked 171 in the list of 500 fastest computers of the world. Amidst the story of all these successes in [20] are also buried references to some major missed opportunities due to bureaucratic delays and misplaced priorities arising from bureaucrats and politicians not grasping technology and its value adequately.

The brief history summarized above offers many lessons for India: (a) India needs to empower its scientists and engineers a lot more than it does today and wrest key decisions away from bureaucratic generalists (like of the IAS cadre) and politicians. (b) The country needs to make investments in key areas with a long haul point of view. (c) Successes that begin to emerge need to be accelerated by active supportive efforts by the government. (d) A consortium approach involving government, industry and educational institutions accelerates progress at incredible rates. Incidentally, these are also the lessons taught by other major efforts like India's nuclear program [21] and the space program [22]. (e) Availability of cheaper solutions from abroad is a double- edged sword, and its non-availability can often be turned into a major opportunity.

Despite these lessons, unfortunately India is yet to empower its scientists and engineers adequately. Some specific steps India could take are: (a) Replicating the successes in the limited areas mentioned above into other areas, and specifically beyond high- tech areas, with the specific view of raising overall quality and productivity as well as the power of Indian industry to compete better internationally. (b) Having an Indian Technology Service of a high caliber of engineers and scientists who can replace the current IAS cadre of generalists effectively in areas pertaining to science and technology. (c) Helping to improve Technology Management Education significantly in quantity and quality to populate the cadre of people needed both in government and industry to lead technology related decisions and technology management.

Despite the major success of India's software and information industry, if one were to examine whether economic liberalization has really helped to accelerate business formation in India of a sustainable kind and induce development inclusive of all segments of the population, then a different picture does indeed emerge. A new focus is needed to change that picture drastically by effecting major paradigm shifts in India's agriculture and many of the classical industrial sectors. Much innovation is needed at all levels to guarantee an equitable distribution of the benefits. There can be no doubt that leveraging the intellectual assets of India is an important key to it. The first step is the realization by all of the need for having more products made in India by Indians and also owned by Indian enterprises.

## The 'Foreign Made' Bug

Some of the responsibility for the sorry state of affairs of India's domestic industry rests also with Indian society at large. The Indian elite are too fascinated by foreign products, and flashing them has become a status symbol. The conspicuous consumption of foreign luxury and fad items like name brand handbags and watches that add little to the nation's economy but only drain its foreign reserves was at one time a turf only of some of the incorrigible super rich. But it has now become a dominant part of even a sizeable part of the middle class.

Indian society today, which has a large number of people born well past India's independence and the early decades of severe hardship as an overhang of colonialism, has hardly any appreciation for the importance of supporting local industries and products and the individual's role and responsibility in that process. The situation has been worsened by the influence of television, movies and advertisements that tout a culture (or the lack of it?) of mindless indulgence. For much larger levels of domestic innovation and new indigenous products, this '*videshi moha*' (the longing for foreign made goods) has to come down. Next to corruption, this is the next most important area in which 'educated' Indians are failing India badly. Indians need to be educated of the difference between a nation that is "balance sheet rich" and one that is only "income statement rich" [23]. If the more affluent Indians would only observe a modicum of national spirit in the way they spend, the nation would develop much faster. As noted earlier, such cultural changes take time. In the meanwhile, India needs to identify strategies that work in the short term also. One of them pertains to a better use of visual media of all sorts.

**Cinema and TV**: One of the major criticisms of the philosopher C.E.M. Joad was how every great invention gets turned by people into a tool for trivial pursuits resulting thereby in a major confusion between ends and means. No other better example of that exists than the obsession of India with cinema and television. No other society is perhaps as involved with these as Indians are. Although these have given employment to a very large number of people in India, countless productive hours are also wasted on these by all age groups. What could be a set of very educational and informational media is debased in India with mindless entertainment and sensational and politicized reports of all types. The government needs to devise ways to induce greater use of these media to inform and educate people (in an unbiased way) on matters that could improve productivity and quality of life. Setting up more channels like those of the Public Television in the USA is highly needed, as also greater support for good documentaries on science and technology.

## Over Reliance on Foreign Consultants & Advice

As though the local craze for foreign products is not enough, we also see many developing nations becoming a victim to various mega projects suggested or thrust upon by foreign multinationals and lending banks with vested interests. These play an organized game of luring client states with project proposals marked by highly underestimated costs and overinflated benefits. Lending institutions collude with such corporations and insist on having many contracts go to the foreign MNCs and not to firms of the borrowing nation. The net result, often, is a high level of indebtedness to the developing country through a foolish effort that yields at many times the projected cost, only a small fraction of what was projected. Several examples of these are given in [24]. Given that India has also become a victim of this game, that book should be a required reading for policy makers and politicians in India.

An example very relevant to India is the Dabhol power project that the Government of Maharashtra entered into with the infamous American corporation, Enron [25].

> "For many on the Asian subcontinent, Enron epitomized the downside of the modern global economic system where powerful corporations from the West often bully their way into development projects that fail to live up to shiny promises.... The lessons of Dabhol may be like the larger lessons of the Enron debacle. A company known for its hubris tried to accomplish too much, too quickly, playing fast and loose with the financial realities and counting on political allies to clear the way." [25]

In Chapter 1, we already noted how some foreign governments rush to provide India some outdated technology at cheaper prices just to starve Indian researchers (who are at the cusp of a major breakthrough) of the support from the Indian government.

Every attempt at securing a product or service from a foreign multi-national comes at the expense of the support for the growth of technology and innovation within the nation. A judicious and honest cost benefit analysis by unbiased specialist teams with significant Indian presence is necessary before embarking on them. The analysis should take into account not only the immediate costs and benefits, but also the long run impact on building capabilities within the nation and the export opportunities associated with them. Needless to say, the lure is not always the cost benefit tradeoff, but underhanded financial and other rewards offered to certain decision makers. Once again, we see that it all leads back to political and official corruption, the greatest curse on India today.

## Defense Expenditure

India is one of the largest arms importers of the world as per the Stockholm International Peace Research Institute (SPIRI) [26]. Unlike China, India has failed to become self reliant and has not created a robust domestic industry in this area. The World Factbook [27] estimates India's defense expenditure to be 2.4% of the country's GDP. But, this could be much higher since, as during the days of the Pokharon experiment (see Chengappa [21]), the Indian government could very well be hiding a substantial chunk of expenses related to strategic initiatives under other more innocuous headings. Reducing the expenditures on defense related imports should be a high priority for the government for both an increased ability to support domestic innovation and businesses as well as for providing employment. In this context, it may become necessary for India to shed some of its ideological stances, like its 'No First Strike' policy, as a cheaper way of deterring repeated misadventures by other countries towards India.

## A Larger Vision

What I describe now is once again a highly schizophrenic aspect of Indian society, and it pertains to how people react to high wealth generating opportunities. On the one hand, we have a class that wants to move ahead to egregious heights through corruption and illegal means, while, at the same time, we have a vast majority, of particularly small businessmen, who are quite content to just get by. Examples of the latter are some fine restaurants that are much sought after in certain localities, and some stores like "Grand Sweets" of Chennai that are famous for their savories, sweets and various spices. Examples can be given from technology areas as well, some highly respected medical testing laboratories being one such set.

In a country like the US, the success of the small establishments will get replicated either by the owners themselves or by investor groups to a much larger national scale, and they will grow into large companies and even spread their wings internationally. A well-known example, known even in India, is Starbucks which started off only "as a single store in Seattle's historic Pike Place Market," got purchased with the help of some local investors by an ex-CEO who got excited by the coffee houses of Italy, and is now an international large enterprise. Domino's Pizza that has scored major success through a product almost unknown in India is another example. One cannot help wonder why such successes do not happen with Indian products and businesses.

Part of the reasons can be attributed to many of the usual problems at work: lack of formalized financial and management skills among owners; unavailability of venture capital or private investment groups; the diversity of India in terms of languages, local customs, and preferences; and the hassle of government regulations and official corruption. But at a fundamental level, society and government have done little to seize opportunities quickly and enthusiastically.

Consider the venerable Indian Institutes of Management. These have certainly helped some large industries of India like Indian textiles. Some can boast today of their graduates in financial engineering being hired at enviable salaries by foreign financial institutions. But they seem to have taken little interest in giving small businessmen like those I described earlier necessary entrepreneurial and management skills, or to usher in a culture of ventures and franchising in India. Nor have they trained a cadre of graduates who can take such unpolished gems and turn them into a dazzling rivière. Again, the 'educated (and educating) Indian' has to bear considerable responsibility for many missed opportunities of this type. The leading educational institutes of India operate, to a large extent, as though their overarching purpose is to win Western approbation, and to get students sent abroad or placed in foreign MNCs. This is a terrible waste of a set of major national resources of very high potential.

There are many department stores in the USA that sell bed sheets made in India of Indian silk cloth. A typical set is priced around $300 in the high-end stores, but it is safe to bet that the Indian's take in that is only a negligible fraction. I recall a time when designer shirts made by "Charagh Din" of Bombay were a rage in India and could be bought in India for the equivalent of $19 approximately. During that same time, in Juan-les-Pins, France similar shirts, but of a much lower quality in cloth and designs, were selling for $100 and up. Why is it that India that is so eager to open its doors so widely to foreign textile chains has taken no steps to take its own wares abroad through storefronts and direct e-marketing?

India is under great pressure, both internal and external, to let in some chain stores from the US into its retail markets. Some of these chains have accumulated a dubious reputation in the US for their predatory practices that have caused much angst among labor and driven many smaller shops out of business. Not only that, when the going gets a bit rough, they close up shop leaving the population nowhere to go for their essential needs. What protects the Indian customer from a similar scenario if it were to play out in India? What does one do with the many displaced small storeowners and street vendors who will lose their livelihood? These are not issues for a debate based on ideology, but critical practical aspects that will determine if India can move on to become a major producer of indigenous products deriving their full benefits, or will simply become a supplier of cheap labor and an irresponsible abdicator of the opportunities offered by its own market.

India has to break the logjam it is in today. In the short run, programs like 'Make in India,' where the primary commodity we offer is labor, may be necessary. But that cannot be for the long term. The lesson coming from China today is that growing one's own internal market and one's own product portfolio is a surer and more stable way to progress, while dependence on foreign sources of funding and over investment in capital assets to support cheap exports are sure prescriptions for economic chaos and downfall. Yet, today is also a time when some learned economic advisors are advocating that India should adopt the very policies that have taught China these bitter lessons [28]. India needs to find a mix that is appropriate for India and not try to mimic any example without a custom refit.

As we roll out a red carpet for others to "Make in India", we should also take inventory of what is already made well in India and help the small businessman scale up and also become global. The model of the co-op handicrafts stores that have helped rural artisans much and improved the quality of handicrafts should be replicated in many of the consumer products that India makes. India needs to find innovative ways of making globalization work for it, instead of against it.

## Standardization

One of the major problems in scaling up the market for Indian products is the absence of standardization that is essential for replication while maintaining predictable high quality and, thereby, meeting customer expectations. Many small businessmen in India, from cooks to craftsmen making handicrafts, work intuitively, but their knowledge and skills are not easily replicable due to lack of standardization. There is no formalized effort at constant improvement of products.

This point can be best illustrated through the following simple example. The lamp shown in Figures 3.2 is a familiar Indian handicraft item that is bought by many foreigners when they travel to India, or can be purchased in some NRI Indian stores abroad. The first challenge faced by the buyer is that the screw size for the bulb holder will not take a holder suitable for one's own country. Also, once the buyer finds an adapter

Figure 3.2 Unstable Lamp

(not easy!) and puts a lampshade on it, the lamp is likely to topple and this could break the bulb. This is because the lamp has a very small base and poor center of gravity. Simply by attaching the lamp to a larger decorative wooden base, or at least by giving a provision for that (two holes in the base and a couple of matching screws) would result in a satisfied customer who is more likely to recommend it to another person. In this instance, it would be unfair to expect the poor Indian craftsman or even the businessman to divine such issues that are seldom faced at home, and it would behoove members of a handicraft council or productivity council of the government to educate them on how to improve the products and make them more appealing. In fact, if every college with an engineering and management department would "adopt" an industry and effect improvements, Indian products would sell a lot more than they do today.

Another example that illustrates the same point is the simple incense stick. Consumers may buy a packet and fall in love with its fragrance. And, say two months later, they may order a pack of the same brand and fragrance line, assuming they will get the same product. However, when they receive the new batch, they may be shocked to find that the new sticks smell nothing like what they had bought earlier. Again, this problem can be attributed to a lack of standardization either in the process of manufacture or in the set of ingredients used to create the product.

Many small players, like makers of handicraft items, lack market intelligence totally, at least, as it pertains to foreign buyers. For example, they still try to peddle the same old brass ashtrays (not knowing how much smoking has decreased in the West) and various idols of Hindu gods to foreign tourists who cannot be persuaded to buy them. Couldn't the government arrange for some market intelligence for these people that would shift their product profile to items that are more secular and will really generate more interest? We, certainly, have great pundits in marketing and branding, but they tend to focus entirely on large corporations. It is equally important they spend their time doing some *pro bono* work for these poor artisans and guide them in ways that will help them earn the money they deserve. In the least, some student projects could be directed to such areas. Once again, there is a need to shift the focus of India's management educational assets, at least to some extent, to the nation's larger needs.

Innovation should not be restricted to the technology sector alone. Even in the USA, the citadel of technology innovation, the real driver of economic progress and employment has always been the small and medium business segments of manufactured products and services [29]. It is extremely heartening to know that in India too entrepreneurship is taking roots at all levels [2].

> "India is being recast, remolded and redefined. ... India is brimming with the aspiration of a billion entrepreneurial minds ... the Indian society is exactly the opposite of what is happening in the West, which despairs that the excesses of capitalism can never be controlled and that the destructive power of inequality is the Achilles heel of capitalism."
> - Sengupta[2]

Unfortunately, the efforts going on in India are not supported by a systematic involvement of government, large industry, and the intelligentsia. New technologies and technology innovations are extremely important for India. But, it is equally important that adequate attention is paid to opportunities for innovation in all areas to ensure not only that economic progress becomes inclusive and does not leave behind those at the lower end of the totem pole, but also helps the country to move forward at a fast pace without being subject to the whims of other nations and nationals.

*\*\**

Nehru, famous for his agnosticism, is supposed to have quipped that one cannot preach God to a starving person. Similarly, one cannot preach innovation, going global, and standardization to the small businessman or worker who is struggling to even get by. Therefore, it is important for the educated and more fortunate people of India to connect with them and their problems, and to bring their own skills, vision, and knowledge to uplift the latter. The stories narrated in [2] demonstrate how those connections and government subsidies can create ripple effects across the entire economy. The 'bottom of the pyramid' in India does not have to be only a big market for MNCs as envisaged by management gurus like the famous C.K. Prahalad [30], but it could indeed be turned into a powerhouse of product creation capturing large markets, if only Indian society could get more connected and could get its institutions to work with an even greater national focus than what they now have.

## SUMMARY & ACTIONABLE ITEMS

**Corruption:** It is the most insidious obstacle to national progress and to domestic innovation, business formation, and economy. India needs special laws at the national level, an independent enforcement authority free from political intervention, and special courts for fast tracking offences. Track the wealth of key government officials and politicians over time and their sources, and bring to book offenders without regard to political affiliation or stardom. Use technology effectively to curb corruption.

**Black Money:** This is the second most serious curse that limits available capital for productive economic use. In addition to implementing solutions similar to those to curb corruption that will also apply to combat black money, India should undertake a major international initiative to force cooperation from countries that serve as safe havens for black money.

**Law Enforcement:** Increase the ability to get justice and that too in a timely manner. Eliminate corruption in law enforcement agencies and the legal system. Mandate time limits for court decisions to occur. Do not allow endless appeals and postponements. Double the penalties when the guilty are those in power. Make the systems work equally for all. Monitor judges on these scores. Subject judges to wealth monitoring, and impose speedy impeachments when they are guilty.

**Passive Assets:** Educate the public about the personal futility and national consequences of having wealth locked up in passive assets like gold. Provide stable, reliable, and honest alternative investment channels backed by the government.

**Market Volatility:** Eliminate fraud and market manipulation. Increase transparency of all registered businesses. Restrict or heavily tax very short-term trades. Curb foreign investors' ability to induce major swings in the market by their speculative actions.

**Social Security:** Increase the ability of the public to participate in the securities market to generate capital for new businesses based on innovation. Provide a safety net to families from catastrophic single events altering their fate. Develop a pension/social-security scheme for all.

**Make Globalization Work:** Leverage India's position as a sought after market to enlarge global markets for Indian products and services, and to get support for India's legitimate international agenda on all aspects like controlling terrorism, bringing back black money, trade and environmental regulations, and seat at important international forums like the UNSC. Do not hesitate to deny uncooperative nations the opening of India's internal markets. Do not empower them more with imports without quid-pro-quo (not as bribes to individuals). Help to create national and global distribution channels for Indian retail goods. Break the barriers that impede the formation of national foot-prints for Indian business.

**Foreign Goods & Services:** Educate the public of individual respon-sibility on using India's foreign exchange reserves wisely. Exercise greater caution with regard to large projects, large foreign procurements, and development loans. Vet foreign consultants and their analyses thor-oughly. Reject contractual constraints that get in the way of participa-tion and growth of domestic industries.

**Defense Expenditure:** Reign in defense expenditure and defense im-ports to divert a larger portion of the GDP towards national develop-ment, innovation, and new business formation. Temper erstwhile ide-alistic stands with a higher dose of pragmatism in matters of strategic importance.

**Aim for Inclusive Growth:** Identify those things that India makes well today in small pockets and help the small businesses to scale na-tionally and globally. Organize support in areas like standardization and product quality improvement. Provide market intelligence and distribution channels for the small businessman. Exploit markets for growth of Indian owned enterprises. Be weary of multinational retail chains. Leverage India's educational institutions like the IIMs and IITs by forcing a greater focus on national development.

**Leverage the Media:** Encourage, and if necessary mandate, greater use of airtime and bandwidth by the media for educational purposes at all levels.

**Social Issues:** Educate parents and teachers to encourage a climate of inquiry and independence. Help foster a climate of constructive up-ward communication resulting in quality improvement and innova-tion. Create a climate for greater participation of the educated middle class in politics and government. Break the nexus of politics and hooliganism. Educate the masses to be less gullible.

**Merit:** Move the country away from the present culture of entitlements based on caste and religion, to a culture of merit and performance. At least, restrict reservations based on caste or religion so that they apply only to the truly economically disadvantaged. Simultaneously, set a definite time frame to eliminate reservations altogether. Instead of lowering standards through reservation schemes, help disadvantaged groups by developing schemes that help them to compete better in a system based on performance and merit.

**Leverage Intellectual Assets:** Create a cadre of Indian Technology Service. Improve in quantity and quality Technology Management Education to populate this cadre and fill key positions in industry. Leverage scientists and engineers in all areas to replicate successes similar to those obtained in nuclear, space, and software technologies. Wrest key decision making power on technology issues away from bureaucrats to reduce delays and for making informed decisions with right priorities.

**One India:** Do not tolerate any rules or practices that hinder the spirit of a "One India." Have zero tolerance towards discrimination based on gender, religion, caste, language, or state of origin. Make it easy for people and goods to move seamlessly across the nation. Create a network infrastructure of product distribution that makes it easy for entrepreneurs to scale the market for their innovative products.

## Notes and References

[1]   Rama Bijapurkar: *We Are Like That Only – Understanding the Logic of Consumer India*, Penguin Books, India, 2007.

[2]   Hindol Sengupta: *Recasting India, How Entrepreneurship is Revolutionizing the World's Largest Democracy*, Palgrave Macmillan Trade, New York, 2014.

[3]   Sean Dennis Cashman: *America in the Gilded Age*, New York University Press, 1984.

[4]   "Another Engineer Found Murdered in Bihar," *The Hindu*, December 29, 2015.
http://www.thehindu.com/news/national/other-states/another-engineer-found-murdered-in-bihar/article8041455.ece

[5]   "Two Forest Officers Stoned to Death in Andhra," *The Indian Express*, December 16, 2013.
http://indianexpress.com/article/news-archive/web/two-forest-officers-stoned-to-death-in-andhra/

[6]   "Desperate, Disgusted but Proud – India's Human Waste Removers," CNN, Oct 2, 2014.
http://www.cnn.com/2014/10/02/world/asia/india-waste-cavengers/

[7]   Jean Drèze & AmartyaSen: *An Uncertain Glory, India and its Contradictions*, Princeton University Press, Princeton & Oxford, 2013.

[8]   "India's Missing Investors," *Bloomberg View*,
http://www.bloombergview.com/articles/2015-04-09/india-stock-rally-needs-more-domestic-retail-investors

[9]   "Indian People Hold 20,000 Tonnes of Gold," *Mining.com*, May 10, 2015. http://www.mining.com/indian-people-hold-20000-tonnes-of-gold/

[10] "Indian Households Hold Over $950 Billion of Gold," *The Economic Times*, Dec 5, 2011.
http://articles.economictimes.indiatimes.com/2011-12-05/news/30474783_1_gold-imports-tonnage-terms-gold-demand

[11] "Government's Gold Bond Scheme Gets Only 8 Crore in First Week," *The Times of India*, Nov 17, 2015.
http://timesofindia.indiatimes.com/business/india-business/

Governments-gold-bond-scheme-gets-only-8-crore-in-first-week/
articleshow/49809904.cms

[12] "India targets tax evaders who hide 'black money' at home and abroad,"
*The Washington Post,* Sep 6, 2015,

https://www.washingtonpost.com/world/asia_pacific/india-targets-
tax-evaders-who-hide-black-money-at-home-and-abroad/2015/09/04/
2532b7c2-50c4-11e5-b225-90edbd49f362_story.html

[13] "Indian black money," Wikipedia,
https://en.wikipedia.org/wiki/Indian_black_money

[14] "Modi's Black Money Hunt Earns Only Enough to Buy India
Icecream," *Livemint e-paper,* Jan 29, 2016.
http://www.livemint.com/Politics/XQLgFn2I1ydEidkulmA38H/
Modis-1-trillion-black-money-hunt-earns-only-enough-to-buy.html

[15] Joseph Stiglitz: *Globalization and its Discontents,* W.W. Norton &
Company, NY, 2002.

[16] Meera Siva: "What's a 'Hindu' Rate of Growth", *The Hindu Business
Line,* June 8,  2013.
http://www.thehindubusinessline.com/portfolio/technically/whats-a-
hindu-rate-of-growth/article4795173.ece

[17] "Crony Capitalism," Wikipedia,
https://en.wikipedia.org/wiki/Crony_capitalism

["Crony capitalism is a term describing an economy in which success
in business depends on close relationships between business people
and government officials. It may be exhibited by favoritism in the
distribution of legal permits, government grants, special tax breaks,
or other forms of state interventionism."]

[18] "Crony Capitalism Hurts the U.S. Economy," *The Wall Street Journal,*
Oct 15, 2015.http://blogs.wsj.com/washwire/2015/10/15/report-
crony-capitalism-hurts-the-u-s-economy/

[19] "India Has Labor, Markets for French Products: Modi", *The Hindu,*
Jan 24, 2016.

http://www.thehindu.com/news/national/french-president-francois-
hollandes-threeday-visit-to-india-modi-hollande-address-indiafrance-
ceo-summit/article8148071.ece

[20] V. Rajaraman: *History of Computing in India 1955-2010*, Supercomputer & Education Research Center, IISc., Bangalore. Published by the IEEE Computer Society. http://www.cbi.umn.edu/ hostedpublications/pdf/Rajaraman_HistComputingIndia.pdf

[21] Raj Chengappa, *Weapons of Peace*, Harpercollins Pub. India Pvt. Ltd., 2000.

[22] A.P.J. Abdul Kalam (with Arun Tiwari): *Wings of Fire, An Autobiography*, Universities Press, 1999.

[23] Thomas J. Stanley & William D. Danko: *The Millionnaire Next Door: The Secrets of America's Rich*, Pocket Books, NY, 1996.

[Based on findings in a research study of the rich for a luxury car maker, Stanley and Danko discovered some shocking and counter-intuitive facts about America's rich. One of them was that they were "balance sheet rich" in the sense of having a significant amount of wealth and being careful in their spending habits, as opposed to "income statement rich" in the sense of being flashy spenders with little wealth but with a high amount of liabilities.]

[24] John Perkins: *Confessions of an Economic Hit Man*, Penguin Group (USA) Inc., 2004.

[This book is by a former 'economic hit man', i.e., a highly paid professional who regularly doctored his findings to help major US corporations to cheat several developing nations. He narrates some alleged real stories and catalogues its devastating consequences on the world and the democratic way of life.]

[25] "Enron's India Disaster," Consortium News.com, https:// consortiumnews.com/2001/123001a.html

[Enron was a US corporation that indulged in a variety of fraud nationally and internationally. Many of its senior executives were indicted and jailed for their role in various nefarious activities. Many pension funds and employees were ruined by these activities in the USA, while untold harm was also done to many of Enron's gullible international clients. Enron enjoyed unquestioned support from the highest levels of US government including President Bill Clinton who used his office to pressure countries, including India, to comply with the corporation's egregious demands].

[26] "India Becomes Top Arms Importer in the World," *Defence Now.* http://www.defencenow.com/news/106/india-becomes-top-arms-importer-in-the-world.html

[27] *The World Factbook,* CIA, https://www.cia.gov/library/publications/the-world-factbook/geos/in.html

[28] ArvindPanagariya: "India's Best Choice Would Be to Emulate China's Export Model," *The Huffington Post,* Jan 20, 2016. http://www.huffingtonpost.in/2016/01/20/india-china-export-policy_n_9028136.html?fb_action_ids=1260172327333172&fb_action_types=og.comments

[29 Peter F. Drucker: *Innovation and Entrepreneurship,* Harper & Row Publishers, Inc., 1986.

[30] C.K. Prahalad: *Fortune at the Bottom of the Pyramid: Eradicating Poverty through Profits,* Pearson Education Inc., Prentice Hall, 2005.

## Chapter 4

# State of India's Educational Systems

Education of a right kind is essential to create and sustain high levels of innovation. There are many books dealing with India's educational system, particularly its deficiencies and how it could be improved. Among the more recent books, two notable ones are those of Drèze and Sen [1], and Kumar [2]. Both of them are good contributions in their own right. This chapter, which is a description of the state of affairs at present, cannot avoid some overlap with the work of these authors. But the reader can expect a considerable departure from the works mentioned both in the assessment of the situation as well as in some of the proposed solutions presented in Chapter 5. Unlike some other authors, I will not shy away from discussing the issue of reservations.

As in the works of other authors, the discussion here will also be centered on formal education in terms of traditional academic curricula. However, two areas not covered presently in schools in any significant and standardized manner, but vitally important for India's progress and strength as a democracy, deserve special mention.

The first and foremost is a formalized curriculum on ethics and morals. Schools play a major role in shaping successive generations and their behavior. For example, teaching in schools the bad effects of smoking has played a significant role in the reduction in tobacco use among the educated population in the USA. Given the high levels of corruption everywhere in India, one cannot overemphasize the importance of incorporating ethics and morals in the school curriculum in a formalized manner from a young age. The secular values of honesty, integrity, and hard work should be taught actively in every school and reinforced *ad nauseam* in behaviors. While these are done in some schools, formalization of a secular curriculum is needed to avoid abuse by any special interest group based on religion or ideology.

The second is the imparting of a good understanding of the democratic process and of a watchful eye that citizens should maintain to guarantee the integrity and efficacy of government. School education should also emphasize to students the importance of a society based on the rule of law. These are the only long-term antidotes to the ongoing sustenance of political corruption, destructive expressions of discontent, and the hoodwinking of vote banks through diversionary handouts by politicians who ignore even basic needs of the people they are supposed to serve.

## 4.1 A Different Optic

Before discussing specifics of improving the educational system, three issues deserve to be addressed: (a) What should be the goals of an appropriate educational system for India? (b) What are the metrics by which one should measure progress? (c) How does one evolve to an efficient system within the various constraints that are indeed real and quite formidable? It is my belief that India will have to evolve its own goals, metrics and solutions, and it would be a mistake to mimic some other country, particularly of the West. A proper filter may also need to be applied to recommendations from those, including those of this author, who have spent much time away from India.

One, however, can agree with Kumar [2] that best practices that have succeeded elsewhere should be considered, and indeed many lessons can be learned from leading foreign universities. Kumar has given a detailed account of how some of the top universities in the US, like Berkeley and Stanford, achieved and continue to maintain their prime status. These institutions have given much to the world. In this context, I must mention *Academic Duty* [3], a book by Donald Kennedy, a former president of Stanford, that is a gem and must be read by every educator.

At the same time, one must note that not all US universities are of the caliber of Stanford or Berkeley. Indeed, there is a wide spectrum in their quality. Many are even getting progressively worse through their increasing use of less qualified adjunct instructors, degrees given on the basis of online education not on par with regular classes, and less stringent admission standards - all in the name of cutting costs and increasing revenue. Lots of gaming does occur in garnering research funding, and there is quite a 'buddy system' in how things work. A very large percentage of published research from the US is of a type that adds little to the field or results in things of no practical value. Most awards and honors of professional societies are highly politicized and managed by committees whose members themselves may have little eminence. There is also casteism based on gender, color, and academic pedigree. The big difference, however, is that the US is a country which, due to its enormous wealth and high inflow of foreign students (who also provide cheap labor as teaching and research assistants and as post-docs), can sustain a shotgun approach to education and research at a mega scale, suffer a large percentage of misses, and yet produce much that is still worthwhile. India is not blessed similarly and has to optimize its minimal resources including funding.

While historically the US has placed significant importance on what is called "liberal education," the same cannot be said for science and engineering education. The current eminence of the US in technical education and research owes much to its first and second generation immigrants to the country as well as to its foreign students and researchers. The cost of education is exorbitant in the US, and increasingly fewer Americans even pursue higher education. Similarly, the society in the US compared to the society in India arguably values higher education less and measures success mostly in terms of money and power. This should not surprise one especially in light of the fact that in the US, due to its business climate and large exports, those without even a college education are able to enter (in honest ways) the richer echelon of the society as owners of businesses, or as entertainers and sportsmen [4], [5].

In short, while India must aspire to achieve the best parts of US education, the question whether India can replicate them without creative adaptation suited to the Indian context is one that raises serious doubts. India must also prioritize its needs and identify a set of need-based expectations from its educated and trained citizens. Chasing some ranking in a western list may not be good, given the cultural bias of those rankings and the high importance given to monetary measures like the amount spent per student. Most importantly, India needs to stem the loss of the crème of its crop to other nations.

Prescriptions like those of Kumar [2] that India should build forty or so mega universities similar to the type that exist in the USA are easily said than done. This is because much of the higher education in India, today, suffers from several key limitations: (a) unavailability of qualified professors; (b) inability to absorb graduates into the Indian work force; (c) too many of top Indian graduates leaving India. What is needed is an unbiased and carefully done Cost Benefit Analysis that is India specific, with a proper accounting of India's gains and losses. I am of the strong view that the type of investment needed to implement the prescriptions will be found to be both ill advised from a cost benefit perspective as well as infeasible.

I am also skeptical of the unconstrained entry into India of foreign venture capital and educational institutions. Business is *ipso facto* motivated by profit, and without precautions in the form of significant equity participation of Indians in India and share of intellectual property, the benefit to India from foreign venture capital could become marginal. Also, if one were to consider the horror stories of some Indian software engineers with regard to their experience with NRI middlemen in the US, it is not clear that employees in India would benefit from NRI owned companies as much as they do as employees in foreign owned multinational corporations of repute. In the least, the government will have to set guidelines and maintain a watchful eye. As noted elsewhere, closely tied with the issue of venture capital is the ownership by Indians in India of the new businesses that will be formed.

With regard to satellite campuses in India of foreign educational institutions, it should not be forgotten that what drives them is a profit motive. There has to be a concerted effort and vigilant oversight to ensure that quality is not compromised, and that the new entrants do follow the curriculum in their parent institutions and impart them with only instructors of the best quality. Distance education is not a panacea either. While distance education using multimedia can certainly augment the resources of the nation, one must note that it is an area where the following words would play out even more strongly than in face-to-face education in a classroom.

> " ... the power of instruction is seldom of much efficacy, except in those happy dispositions where it is almost superfluous." – *Edward Gibbon.*

Turning to Drèze and Sen [1], they are focused a lot on the lower echelons of society whose uplifting appears to be their main concern. They even lament about the "success of the first boys" in India, the "first boys" referring to the very few who are able to even get higher education in a top institution. While I share deeply the concerns of Drèze and Sen for uplifting the poor masses, given my emphasis on innovation, I feel that it would be a bad mistake for India not to consciously nurture a set of "first boys (and girls)." However, that set must be based only on the criteria of merit, and academic and creative excellence, and formed within a system that provides a level playing field for all. Except for this, I agree with the authors' following assertion that is justified by the litany of deficiencies listed in their book.

> "... despite the great success of the first boys, India's education system is tremendously negligent both in coverage and quality. The steep hierarchy that has come to be tolerated in India is not only terribly unjust, but also extraordinarily inefficient in generating the basis of a dynamic economy and a progressive society." - *Drèze & Sen* [1].

The lists of deficiencies cited in [1] and [2] are embarrassingly long. A detailed discussion of these deficiencies can be found in those books. Here, only a brief listing and discussion of the most important ones are provided. It certainly appears that these deficiencies are a result of India, at a macro level, not having given enough attention to some fundamental issues related to goals, metrics, priorities, and management of education.

The deficiencies, however, do not make me despondent. Based on my experience as a student and faculty in India, as well as my extensive interaction over the last several decades with Indian academics, I can assert without any reservation that despite some (correctible) lapses in the area of hands on practice and a few others like a clear understanding of what constitutes quality research, there is no dearth of highly knowledgeable or inspired and inspiring academics in India. Finally, as a researcher who has straddled both theory and practice, I aver that without the backing of a good theoretical foundation – the forte of Indian higher education - it is hard for one to pick anything but low hanging fruits. Even more importantly, it is much easier for a theorist to move towards practice than vice versa, and India does indeed have a sizable pool of those well versed in theory.

In short, there is a need to temper one's frustrations with a genuine empathy for the constraints within which Indian academia has been forced to operate and also to have an appreciation of its latent potential. One's views cannot be driven only by success elsewhere that is owed to a lot of other factors besides one's own personal drive or merit. Unlike many other parts of the developing world, India is capable of making major improvements on its own and may need only a small amount of help from outside. The solution lies in the recognition of the need for a major change and the empowering of the right people within the system.     None of this is, however, a call to ignore the true deficiencies in India's educational system to which we turn now.

## 4.2 Deficiencies of the Indian Educational System

Independent India's accomplishments in the area of education are certainly phenomenal when judged in the context of the challenges posed by where the country started, the explosive growth of its population, pressing priorities related to poverty alleviation, and the tremendous pressure for defense preparedness imposed by some hostile neighbors. See [6] for a detailed set of tables from which the statistics reported here are gathered. In 1951 (just four years after India's independence and one year after India became a Republic), literacy rates were only 8.9% for females, 27.2% for males, and 18.3% overall. They increased to 64.6%, 80.9% and 73% respectively in 2011. This is no small an achievement given that it belongs to a period in which, per the mid-year estimates of the US Census Bureau's International Database, the Indian population saw a phenomenal increase from 369.8 million to 1,189.2 million. Table 4.1, provides a comparison of several other metrics attesting further to that phenomenal progress in school education covering a variety of areas: number of schools, government outlay for education, general enrollment ratios, female participation both as pupils and as teachers, and, reduction in drop-out rates.

Unfortunately, India's achievements fall far short when snapshot comparisons are made to international standards and benchmarks. Also, the numbers in Table 4.1 belie some of the deeper problems related to quality. We list in Section 4.3 the major defects of India's schools. It goes without saying that there are certainly some institutions that are oases of excellence, but they are too few in number in the vast and arid desert of Indian education and quite unaffordable to most Indians.

When it comes to higher education, the diversity of India is even more evident and appalling. On the one hand, Indian graduates have spread their wings around the world to an extent that they are considered a threat to local labor in many countries. Consider the following quotation reflecting the paranoia and xenophobia in the US, a country whose industry is pushing the envelope in many new areas like energy, environment, automatic and intelligent vehicle systems, cell based therapies, and the like while delegating much of the grunge work to outsiders primarily to save cost.

> "Leading American newspapers have carried articles urging improvements of the education and training in the USA, in order to keep up with that learned lot from distant Asia, In-

## Table 4.1 Comparison over the Years

|  | 1951-52 | 2000-01 | 2011-12 |
|---|---|---|---|
| **Number of Recognized Institutions:** | | | |
| Primary: Grades I-V | 2097 | 6387 | 7143 |
| Upper Primary: Grades VI-VIII | 136 | 2063 | 4788 |
| Secondary: IX-X | NA | 877 | 1283 |
| Senior Secondary: Grades XI-XII | 74 | 384 | 841 |
| **Expenditure** | | | |
| Total expenditure in million rupees | 644.6 | 824,864.8 | 3,511,457.8 |
| As percentage of GDP at current price | 0.64 | 4.28 | 4.18 |
| **Enrollment Statistics (in millions)** | | | |
| Primary: Grades I-V | 13.8 | 64.0 | 72.6 |
| Upper Primary: Grades VI-VIII | 2.6 | 25.3 | 33.1 |
| Secondary: IX-X | NA | 19.0 | 34.1 |
| Senior Secondary: Grades XI-XII | 1.5 | 9.9 | 21.0 |
| **GER (Ratio of Enrolled to Eligible by Age)** | | | |
| Primary: Grades I-V | 42.6 | 95.7 | 106.5 |
| Upper Primary: Grades VI-VIII | 12.7 | 58.6 | 82 |
| Grades IX-XII | NA | 33.3 | 56.8 |
| **Ratio of Female Students to Male Students** | | | |
| Primary: Grades I-V | 0.41 | 0.82 | 1.0 |
| Upper Primary: Grades VI-VIII | 0.22 | 0.75 | 1.0 |
| Secondary: IX-X | NA | NA | 0.93 |
| Senior Secondary: Grades XI-XII | NA | NA | 0.93 |
| **Ratio of Female Teachers to Male Teachers** | | | |
| Primary: Grades I-V | 0.20 | 0.55 | 0.79 |
| Upper Primary: Grades VI-VIII | 0.18 | 0.62 | 0.76 |
| Secondary: IX-X | NA | 0.54 | 0.66 |
| Senior Secondary: Grades XI-XII | 0.19 | 0.42 | 0.66 |
| **Total Drop-out Rates** | | | |
| Classes I-VIII | NA | 53.7 | 40.8 |
| Classes I-X | NA | 68.6 | 50.3 |

dia included, who are .... keen on snatching good employment opportunities away from simple Americans." – *Drèze & Sen* [1].

Yet, except in a very small number of areas where the Indian government has placed high emphasis, India has little to show as real prowess in terms of domestic innovation, new products, formation of companies, or capture of a high share of the global market. Not only does the (still small) number of Indians shining abroad belie this sad fact, but it has also added to the problem through what is described rightly as "brain drain," particularly of the crème of India's crop. Certainly, the deficiency of the Indian educational system is a major contributor to that sad state of affairs. It is not able to provide aspiring scholars opportunity at higher levels of university education. Nor has it taken an adequate initiative to lure back an adequate number of NRI specialists and educators of the right type who can meet its needs.

Unfortunately, in official and government circles and even in the society at large, a good understanding does not exist widely of the need to change. One reason is the recent success of the Indian software industry that has added much confusion to the scene, leading the uninformed Indian to believe that all is well. This has been bolstered further by the large "middle class" it has helped to create in India, although it blissfully ignores the fact that the nomenclature "middle class" is based on Indian conditions and would not pass in the USA where the defining characteristics are things like "a single family home with a two car garage." An unassailable consensus is, however, emerging within Indian education circles and leading industrial organizations that India's education system needs to be thoroughly overhauled.

The deficiencies of India's higher education system are examined briefly in Section 4.4. The discussion here is centered only on the main issues. Let us begin with education in India's schools.

## 4.3 Primary and Secondary Education

Overall, India has approximately 1,426,000 schools of which 791,000 are primary (grades I-V), 401,000 upper primary (grades VI-VIII), 131,000 secondary (grades IX-X), and 103,000 senior secondary (grades XI-XII) schools [6]. Yet, compared to the size of the population, the available number of well-staffed schools is significantly inadequate. For example, 12% of the schools operate as one-teacher schools either by design or due to teacher absenteeism. The overall pupil/teacher ratio is high and in the range of 23-40 depending on school and grade level.

**Functionality:** Drèze and Sen [1] assert that "the functioning of the schools remains seriously - perhaps even disastrously - deficient." In many rural and government run schools, absenteeism runs high, at the rate of 20% for teachers and 33% for students, reducing the effective school days to 50 in a year.

**Quality of Education**: Children learn little in most schools, particularly, in government run schools. Teaching methods are quite often based on "mindless rote learning, including repetition - without comprehension - of what has been read, and endless chanting of multiplication and other tables" [1]. Instruction is unidirectional with hardly any practical, laboratory, or hands-on experience, group work, or exploratory self-learning.

**Regional Disparities:** Significant regional differences exist. This is reflected in the highly disparate levels of student proficiency in the nation across states. As an example, consider the comparison from [1], given below in Table 4.2, of children in the age groups 8-11 for two states, Uttar Pradesh (UP) and Kerala.

**Private Schools**: Most private schools and private colleges are operated for profit and often by politicians and their relatives. They operate

**Table 4.2: Proficiency Levels for Children 8-11**

|             | UP  | Kerala |
| ----------- | --- | ------ |
| Reading     | 29% | 80%    |
| Writing     | 51% | 77%    |
| Subtraction | 22% | 64%    |

as "extractive money making machines [1]", and their accessibility is severely limited by geographic location and economic affordability. They extract significant capitation and other fees despite laws prohibiting them, sometimes with devastating consequences [7]. This is even more pronounced in the area of higher education at the college level. While a few private institutions do maintain a high standard, most are substandard in facilities, staff, and instruction. Like in the US, the few better ones steal away good students from the public system and perpetuate a culture of elitism and snobbery.

**Teacher Quality**: Like everywhere else, except for a few driven by their idealism, the teaching profession in India too does not attract the best of the lot, primarily due to lower salaries and worse working conditions compared to jobs in industry and in government - the latter, of course, taking into account extra-legal income. Nevertheless, as noted in Drèze and Sen [1], one could argue that teacher salaries in government run schools are decent when compared to the annual per capita GDP. The real problem is that many teacher appointments are based on nepotism, caste, and bribe, with merit and aptitude playing a minimal role. That has turned the teaching profession in India even more into a last resort of the least employable. As for private schools, except for some elite ones, the average teacher salaries are lower than what they are in government schools.

Drèze and Sen [1] argue that teacher salaries in India are many times above the per capita GDP and exorbitantly high compared to the income of the agricultural laborer. There is an implied suggestion to match the income of teachers to that of the population they serve. If one extends this argument also to other professionals like doctors, the poor will only be in worse shape as they would get worse quality of service than what they presently do. For this reason, I do not agree with that argument at all. The need is to attract the best and the most talented people into the teaching profession and to extract the greatest benefits from them. If anything, it calls for increasing teacher salaries. The real need is also to weed out corruption in appointments. Systems should, in addition, weed out the uncommitted and the underperformers in the profession, and provide periodical professional development for those kept in the system. The power of the unions in protecting incompetence and poor performance should be curtailed significantly (in education and elsewhere).

**Evaluation & Assessment:** There are some major gaps in evaluation and assessment processes. Unlike in developed countries, there are no standards even at state level against which schools and students are checked continuously through standardized tests with a feedback loop identifying specific areas for improvement at the individual student, class, and school levels. The lack of such continuous feedback and course corrections results in a situation of flying blind. The Right to Education Act of 2010 made things worse by creating a set of bad rules including the automatic promotion of students. Most schools operate as though their central goal is just to manage the passing percentage of students in exams at certain specific grades and to get them a certificate.

**Facilities:** A large percentage of schools lack the basic facilities of a playground, gym, laboratory, or library. Many are run in make-shift shacks and overcrowded facilities that flout even basic standards of safety. Adequate classroom materials are not available for a vast majority of school going children. Many schools are marked by an unclean and depressing environment both for pupils and teachers.

**Exam Orientation:** Even the better schools operate with a system of instruction driven primarily by credentialing examinations of state or central governments given at the highest grade levels only. Almost no attention is given to overall personality development, comprehension, and intellectual growth. The hyper-competition for admission into better schools and later into colleges and universities puts too much pressure on pupils to score high marks in the final examinations. That leaves little time for extra-curricular activities, robs the young of a normal and healthy childhood, and also results in much student burn out.

**Special Needs**: Special education for the handicapped or learning impaired is a luxury available only in a select few private schools in urban areas. Traditional services of a counselor or a nurse are similarly unavailable in schools.

**Student Experience:** Overall student experience is poor. Many schools, including privately run ones for the rich and the elite, still mete out corporal punishments even for minor infractions. Students and parents have no recourse in an environment where demand far exceeds supply and the risk of being kicked out looms high. Those who can afford, including those with children enrolled in expensive private schools, have to rely on private tutoring, which in some instances, cost, according to [2], as much as a third of the family income for many middle class parents.

*\*\*\**

The deficiencies described above are truly embarrassing. They drag down India immeasurably. Many reasons have been cited for this sorry state of affairs. At a high level, education in India is a "state subject," that is, by and large, controlled by individual state governments within their territories. The vast differences in the economic and social conditions prevailing in individual states also affect the schools' access to resources. Added to this are the problems of bad government bureaucracies and school managements with a whole host of ills at all levels – politics, nepotism, corruption, permanency of tenure, lack of accountability, seniority based promotions, and absence of performance based rewards.

Even states once considered progressive and enviable, particularly in the area of education, have not escaped the progressive deterioration in quality [8] [9]. Politicians and bureaucrats, themselves lacking in education, an appreciation for its value, and, above all, skills and training relevant to managing educational systems, can be cited to be the singular cause of the sad state of Indian education. If they stopped meddling with education and educational institutions and allowed real educators and specialists to step in and improve the systems, India has certainly the potential to change fast in many noticeable ways.

## 4.4 Higher Education

Catering to India's need for higher education – i.e., "university education" as it is called in the West - is a host of universities with affiliated colleges in the style of UK, a set of institutes, and various specialized centers embedded in certain government research organizations like the Bhabha Atomic Research Center. It is no accident that most of the finest educational institutions in India are predominantly those created after India's independence and shielded from various political pressures to compromise on merit and quality.

As for the universities, they were started in 1857 by the British with three universities (Bombay, Calcutta, and Madras) and their 28 affiliated colleges. The university system has now mushroomed to 712 universities in 2014 (of which about 200 are private), and to a staggering set of 36,671 colleges [6]. Of the universities, except for about 50 launched by the Central Government, the rest were started and are being run by the state governments. The universities operate under the University Grants Commission (UGC) set up in 1956, which over the years has, unfortunately, gained a reputation as an obstructer rather than as a facilitator of progress [2].

Besides the universities, the country has 16 Indian Institutes of Technology (IIT), 19 Indian Institutes of Management (IIM), and 5 Indian Institutes of Science Education and Research. These are some of the gems of India. In addition, India is home to several world renowned institutions like the Indian Institute of Science (IISc.), Tata Institute of Fundamental Research (TIFR), the Indian Statistical Institute (ISI), the Indian Council of Medical Research (ICMR), and a whole host of others specializing in various areas from social development to agriculture. There are also about 40 laboratories of the Council of Scientific and Industrial Research (CSIR) engaged in applied research. This entire system caters to more than 30 million students, a projected number for 2012-13 based on figures in [6], with some of the research institutes not taking any students at all.

Barring some of the elite institutes, within most other educational institutions quality varies from what is passable to the very despicable. The universities and affiliated colleges have remained the most abysmal; unfortunately, they account for a very large percentage (86%) of the student body [2]. The following list of specific deficiencies is given here mainly with a bias towards innovation and to focus our discussion in the next chapter on possible avenues for improvement. To keep the length of the discussion within reasonable limits, we concentrate only on the most important ones impacting innovation.

## The Major Gaps in Higher Education

*Deja-vu:* Many of the shortcomings we noted in India's school education are seen also in higher education in India. Specifically, these include the following: lack of uniformity in terms of quality; inadequate quantity and quality of teaching, professors, and facilities in a large number of colleges; rampant corruption and nepotism; proliferation of private institutions (many owned by politicians) which operate extracting egregious amounts of money without providing commensurate quality of education; reservations based on caste and religion; and complete disregard for the importance of merit in matters relating to appointments and admissions. There is no need to elaborate on these once again.

**Oases, Too Small:** The top institutes of India are certainly like the few oases in a vast and arid desert, and they are able to meet only a small percentage of the demand from even high performers. For example, the combined enrollment of all the IITs put together accounts for only 0.2% of the enrollment in higher education, and only 2% of all engineering enrollment; these figures are based on data for 2011 quoted in [2].

**Academic Silos:** In almost all institutions, including in the elite institutes, the organization of education is within silos organized according to narrow disciplines. Course requirements for degrees are too rigid and do not give adequate choices to the student to enlarge one's ambit into related fields through optional courses, projects, and the like. Dual degrees in parallel are unheard of. Systems totally ignore

the increasing overlap and the merging of boundaries that have occurred in the sciences and in engineering disciplines. As an example,
some new fields like bio-medical engineering are really the splicing of
several fields of science and engineering together and do not lend themselves to a silo based approach. Even the classical field of mathematics
has come to have significant overlap with other areas like computer
science and computing. Many major innovations around the globe
occur in such fields that cut across several academic disciplines.

The present organization in India of faculty and courses for degrees
does not support inter-disciplinary skill development that is very essential for innovation in many high-tech areas. Hiring of faculty also reflects a silo mentality.  For appointment as faculty in any silo, it is
considered absolutely essential to have a degree in the specific subject
area of that silo, without regard to related expertise that should be
represented strongly, or the accomplishments of the applicant in closely related subjects or even in the subject of the silo. (When I tried for a
position in India, this was a serious impediment with engineering departments - despite the fact that one of the largest companies and a
most prestigious research laboratory in the US had no problem en-
trusting me with some serious engineering problems or that the US
university engineering departments trying to recruit me could not care
about the specific discipline of my doctoral degree).  This results in a
situation where supporting subjects are taught by faculty in other departments who have no domain knowledge in the field represented by
the silo, or by those in the silo who are not real specialists in the subject
area. Very little research interactions and collaborations occur across
silos to the extent that it is rare even to find faculty from one silo in
another silo's seminars on topics of common interest. The latter contributes particularly to the staleness of research and disconnect from

the practical for many of the 'pure' sciences. For engineering departments, it inhibits needed cross-fertilization of ideas and collaboration of people.

**Syllabus and Instruction:** In most universities and private institutes, the syllabus is antiquated. Faculty members, including those teaching masters level courses, have no research or practical experience. In many places, a good library does not exist, and state of the art textbooks and journals are unavailable. Even when they are available, they are seldom used. The preference is for books by domestic authors that are more tailored to preparing one to pass the year-end or semester examinations. Even masters level students never read journal articles to get a taste of research from first hand reports. Instruction is mostly unidirectional from the lecturer to the student without adequate opportunity to explore and learn on one's own or in teams. As questions are not actively encouraged, most students take in the lectures passively without interacting with the lecturer. Practical hands on work, which enhances understanding and prepares one better for applied research or the job market, is non-existent or minimal. Open-ended questions and discussion of vaguely formulated problems (as those in real life) are seldom tried out, and that reduces education to a repetitive, regurgitative process. The overall attempt to satisfy the middle of the class leaves the brighter and creative students inadequately challenged. The following comment about an institution that caters to the crème of the crème of students in India is indeed shocking.

> " ... during these four years at IIT, I felt underwhelmed by the academics and faculty members. Barring a handful of professors, no one challenged us intellectually ... There was minimal focus on building things." - Kumar [2].

**Soft Skills**: Internship or co-op programs in industry are rarely available to students. No systematic effort is made to increase personality skills and personal discipline such as being able to effectively manage one's time and to be a good team player. Presentation skills are very poor, and it is common to see the use in conferences, even by some faculty members, of unreadable slides reproducing pages straight from the papers. Presentations themselves get the listener lost in details losing sight of the forest for the trees. Soft skills are, after all, necessary for one to succeed in the job market and in the profession that one enters. Indeed, by one assessment as many as 60-90% of the graduating students in India are considered unemployable. As a result, many major corporations in India are forced to have their own extensive training programs for new employees. That diverts a sizable amount of their outlays away from research and development or even from advanced technical training that could benefit innovation.

**Theory Bias:** Overall, in Indian academia, there is a strong bias towards theory as opposed to the applied and practical. This is due to several factors: absence of faculty with practical industry experience; unwillingness or inability to hire adjunct lecturers from industry; limited availability of resources like laboratories and equipment to support applied research; the lack of industry-university collaboration; a preponderance of returning foreign trained scientists being theorists for whom job opportunities are inadequate in foreign nations; and most of the celebrities and heroes of Indian science hailed by the society being theorists. In certain academic areas, elitism of the purists has even killed progress in India by relegating a lower status in the pecking order to researchers outside academia and things applied. It is not rare in India for someone like me from industry to get a left-handed compliment, "Oh, I didn't realize you do theoretical work too," that would pass for an insult in the USA. In short, the overwhelming attitude is that theoretical and applied knowledge belong to two different worlds, and the twain shall never co-exist, at least in academia. One wonders whether this overwhelming tilt towards theory and elitism is what accounts for the sad phenomenon reflected in the following question.

> "Why did Berkeley faculty admire and love the IIT students, but most of the IIT faculty members feel the same students were not fit for research?" – Kumar [2].

**Faculty:** The majority of the faculty, even in the best of engineering institutes and colleges, lack industry experience. In addition, as noted earlier, in the universities and private colleges, most faculty teaching masters level programs do not have any research experience. Unlike in the USA and other places, the faculty put in little effort in constantly upgrading their skills or in staying current by taking courses. Sitting in lectures of another faculty member who teaches a course in a newer area would be considered absolutely unthinkable for senior faculty in the Indian system steeped in hierarchy and rank consciousness.

**Metrics of Success:** Except for a small number engaged in strategic work of the government, the more research oriented and competent faculty in India are, unfortunately, disconnected from major national priorities and pursue a research agenda aimed primarily at earning personal recognition and approbation from abroad through journal publications. They hardly make any effort to generate innovative ideas of commercial value. Efforts to create an infrastructure to support significant generation of patents and other intellectual property that could be commercialized and licensed are just starting but are too small compared to the needs and possibilities. Most notably, most systems do not permit faculty to get monetary rewards through inventions, thereby taking away a significant catalyst that could motivate innovative efforts. Extra-academic engagements like industry consulting and earnings are even forbidden by most institutions. Academic salaries in

India are significantly less compared to industry salaries and abysmal when compared to what one can earn abroad. That contributes to a flight of the best out of the country or into industry (often for jobs far below their capabilities and talent). The net result is a schizophrenic faculty composition comprised of a few idealistic and very good (and often reclusive) researchers and a majority of those forced to take refuge in academia since it is the only place that would hire them. Although there are some notable exceptions, to a considerable extent this sort of a selection bias is seen also in the PhDs who return from abroad and enter Indian academia.

**The Bureaucracy:** Much control in the government, both at the central and state levels, rests with bureaucrats who have no real skills to make meaningful rules for education or to administer educational systems. In the areas of science and engineering in particular, the appointees to various key positions in the central and state ministries are from the Indian Administrative Service cadres. This cadre of generalists with little specialization cannot be expected to do well in specialized areas. Furthermore, though once comprised of nothing but the brightest of India, it has come to be emaciated too through lowering of standards and rampant corruption, making it often a bane outdone only by the politicians who appoint them to various positions. Certain organizations like the University Grants Commission (UGC) with controlling authority also operate in antiquated and bureaucratic ways stifling progress; see [2] for several examples such as a highly ill advised direction by the UGC in 2014 to the IISc to stop its coveted four year undergraduate program. Even less meaningful is its mandate that undergraduate institutions should not engage in research. The bureaucratic officials micromanage almost every aspect of education, from determining the salaries and qualifications (often with a one shoe fits all approach) to square footage of space allocated to each student. That leaves no discretion to those who can effect improvements. In addition, all around, key appointments like those of Vice Chancellors of universities are highly politicized. In fact, politicization of education and corruption have degraded and continue to degrade Indian higher education significantly.

**License Raj**: Kumar [2] rightly notes the prevailing license raj in Indian education whereby almost all things need government approval. Nothing is free from government interference, and no university can be started without an act of Parliament. The leaders of the nation that helped to set India free seem to have fettered its education through such constraints instead of having government play just a constructive, supportive, supervisory, and monitoring/accrediting role.

**Research/Teaching Dichotomy:** Higher education in India suffers from a dichotomy between teaching and research. To understand the falsity of that dichotomy, we refer the reader to P.J. Hilton [10]. However, with neither a strong research focus nor an industry focus among faculty and in the training of students, most educational institutions in India serve mostly a credentialing purpose through a degree or a certificate that is highly inadequate for a career either in research or in industry. A well-designed system should have the capability to train both types of students, the ones that wish to pursue a research career and the ones that want to move into industry, by enabling them to make choices based on their aptitude and goals.

**Research Quality:** India can boast of an extremely large number of PhDs. The IITs alone are producing around 1,000 PhD scholars per year in engineering and technology. A recent committee headed by the noted Indian scientist, Kakodkar, has fixed a target of 10,000 PhD scholars per year [11] in the engineering sciences alone. But, will the number of PhDs alone matter if a vast majority of them are not of reasonable quality and cannot generate good research output?

With rare exceptions, the best research in India occurs mostly in its elite institutes. Outside of these, what passes as (even doctoral) research in science and engineering is mostly of a type that would not be acceptable to any good Western university or to any of India's leading research institutions. Many research efforts resemble graduate student projects in the US. They tend to be mostly laborious minor extensions of the work of others and significantly lack in originality. In many places, it appears as though no one asks the three critical questions leading to good research: "Why should one do this?", "Is this worth X years of my life?" and "So what?" An overwhelming majority of the publications from India, therefore, ends up in sink journals or in conferences that are organized primarily as the only source of outlets for such work. Often, people are extolled and rewarded for the sheer number of their publications without regard to quality or true impact. Worse still, these PhDs who enter academia create yet another generation of PhDs of still lower quality. With all due respect to those [11] calling for a significant increase in the number of PhDs in India, I would beg that greater attention be paid first to improving the quality of the Indian PhDs and research output. Developing a well defined set of criteria to assess the quality of PhD dissertations and assessing what fraction passes a meaningful threshold alone would constitute a major step in paving the way for improvement. Research has to be highly privileged and yet subjected to a very high level of performance expectations.

**Conference Shenanigans**: Much gaming occurs, particularly in small private universities and institutes, to garner research funding and accreditation. One of them is the preponderance of conferences featuring many foreign researchers as invited speakers, and little credible research work that is locally generated. In many of these, separate sessions are held for local researchers from India just to satisfy some funding agency requirement on the attendee. The serious researchers at the conference do not usually attend these sessions, and speakers in them do not even get the benefit of some constructive feedback from the experts.

The total expenditure of the Indian government for higher education and research is large, but much of it is misdirected and misused. This clearly exemplifies the fact that the solution does not lie in throwing more money and increasing some raw numbers, but in making sure that money finds the right recipients and brings back well-defined and targeted results. Also, serious and credible scientists and engineers need to become involved in decisions relating to funding research and evaluating proposals and completed research. Such an environment would benefit many other efforts like the accreditation of institutions that teach at the masters and PhD levels and in finding constructive ways of improving the overall quality of research and scholarship.

## 4.5 Campus Politics & Indiscipline

Many Indian campuses are marked by too much politicization of both faculty and students. An academic campus should be a place of learning, exploration of new knowledge, and the shaping of future generations as productive citizens and critical and original thinkers. But, there should be no tolerance for politicization of the campuses or for any acts of disruptive demonstrations or hooliganism. Certainly, the students who are the future citizens of the land need to be made aware of all types of legal, economic, and political thoughts and should be free to examine them critically. But those activities need to maintain primarily an academic tone and should be conducted with proper decorum in appropriate forums like classrooms and seminars. Academic freedom does not equate with freedom to disrupt the working of the institution or to indulge in disruptive demonstrations that disrupt society at large.

In almost all instances, the troublemakers in Indian campuses are from the humanities areas. This is not an accidental coincidence, for, in India, except for a minority who enroll in these areas out of a genuine interest for the subject and a career path in mind, the vast majority end up in them due to their poorer scholastic performance and inability to get into science and technical programs. The problem is further compounded by policies that admit students with low academic qualifications to fill reserved quotas. While the technical and private institutions in India have done a reasonably good job of keeping such politics out of their campuses, the government run universities, unfortunately, have had a poor record. In the interest of better education for all, it is absolutely necessary for the government to crack down on political demonstrations of all sorts in educational campuses. Since prevention is always better than cure, it behooves the government to revisit the issue of reservations and admit students based only on merit and a strong desire to learn and get into a career. That holds for faculty appointments as well. The politicians of the land should not use students for their nefarious purposes and should not turn every mishap or legitimate disciplinary action in a campus into a political drama. The taxpayer who subsidizes education in these institutions deserves a better deal from them as well as from the students and faculty.

The present highly charged political environment in campuses does not support the creation in India of mega universities following the American model. Presently, students in the engineering and other specialties in India are spared much of the politics and are able to engage in their studies seriously without unnecessary distractions. Until such time that India can get a handle on discipline and law and order in its campuses, and particularly in the humanities sections, it would be foolish to make the campuses very large and to risk the education of those who will make some genuine contributions to the Indian economy and to India's progress.

## 4.6 The Case for an India Specific Approach

India's real challenge is to ensure that its investment in education, particularly higher education, pays back significant returns, and benefits India directly. At lower levels of education, that involves uplifting the masses and turning them into informed, productive citizens enjoying a good standard of living and turning out high quality products and services. At higher levels, that involves, in addition, the enabling of the graduates to contribute in ways that will help India meet its critical needs and priorities, and enable India to compete more effectively in the global spheres, and ascend as a world leader in arts, literature, science, and technology. India should leverage its education system to accelerate India's move towards greater prosperity and stature, and for the nation to catapult itself into the set of developed nations of the world.

In this, India's record has been terrible. Poverty is still widely extant, and the standard of living is very low for too large a number of people. At lower levels, many have been excluded from real education, and a large percentage of those included have not benefited adequately from their education. At higher levels, while there may be pockets of excellence in a small number of fields and institutions, overall, India lags badly with respect to the rest of the world in its scientific and technical research, and even more importantly, in new product generation. It may also be argued that much of India's investment in higher science and technical education has contributed more to certain developed countries than to India. Changing this sordid narrative should assume a high priority for the nation.

Daunting as the challenge may appear at first sight, we can be reassured by the fact that we are dealing with a nation that has conquered impossible odds in areas like food sufficiency, nuclear and space engineering, and defense preparedness. The absence of a widespread recognition, particularly in the political and bureaucratic classes, of the need for improvement measured by the right metrics is what holds things back in Indian education. India can and needs to craft its own solutions and implement them with a high level of seriousness. There is a need to really redesign India's educational systems and institutions with an emphasis on national priorities, the need for skilled workers in various occupations and ranks, and the ability of the market to absorb graduating students into the labor force.

The polity needs to recognize and be made aware that while a democratic state owes equal opportunity to all its citizens and should provide a large helping hand to its disadvantaged citizens, it cannot afford to compromise merit and performance in that process. Nor can a country like India afford to continue an entitlement system that yields benefits only to a few. A highly competent work force at all levels is a *sine qua non* for the nation's progress in every sphere, be it construction, health, manufacturing, science, or technology. The country also needs such a force for building durable, efficient, and reliable systems and to provide service of high quality. Restoring the emphasis on merit is absolutely necessary to enhance quality of life for all segments of the population including the poor. What is needed is a commitment to merit, with a simultaneous commitment and concerted effort to creating a level playing field of competition, and a conscious attempt to lessen the wide disparities of income and dignity.

Thus, some "sacred cows" like reservations based on caste and tribe without additional strings need to be sacrificed. The last many decades have shown that the uncritical way the reservation system has been perpetuated has only enriched a select few, while leaving the country with a substandard set of administrators and workers in many walks of life. The most impacted are, once again, people at the lower end of the economic pyramid. The more affluent can and do buy quality services. But the poor are increasingly unable to buy at prevailing exorbitant prices quality services, even in such critical areas like health care. If a metric of success is the availability and easy access of essential services and goods in near equal quality for all segments of the population, it must be admitted that the reservations policies of India have been counterproductive and have been a dismal failure.

The above comments are not meant to ignore any past or current inequities and injustice induced by social factors like caste. Nor is it an attempt to turn back the clock. It is just a well-deserved indictment of the abuse of a system made for some great ends. Not only do I acknowledge the injustice, but I will go even a step further and advocate strongly that the government, and society in general, should device systems to help the disadvantaged to gain the appropriate level of merit to avail opportunities at all levels. We should also create an environment of zero tolerance for perpetuating the discriminatory practices of the past. We consider it entirely possible for India to provide a much better quality of education and better standard of living for all, while maintaining an emphasis on high quality and performance. That, however, requires a strong commitment on the part of all segments of the government and the populace, and a resetting of the metrics of success to reflect true advancement of the nation. Naturally, the current culture of entitlement must be replaced by a culture of merit-based rewards along an achievement scale reflecting the needs of the nation and the society. It calls for a redefinition of the very term 'disadvantaged.'

With regard to higher education in India and the literature on Indian higher education, the emphasis on customary measures used in the West may also be quite misplaced. When one gets under the hood and examines the way researchers and educational institutions are ranked in the West, one will note that the metrics include many variables that are not easily quantifiable and many that may not even be relevant to present India. As an example, consider the deficiencies of the various citation statistics and impact factors so commonly used and touted. Upon careful analysis, a Joint Committee on Quantitative Assessment of Research of some leading scientific organizations and made up of several highly respected members of academia who are also specialists in quantitative sciences concluded as follows.

> "The accuracy of these metrics is illusory.... The sole reliance on citation data provides at best an incomplete and often shallow understanding of research – an understanding that is valid only when reinforced by other judgments. Numbers are not inherently superior to sound judgments. .... Using an impact factor of a journal alone to judge a journal is like using weight alone to judge a person's health. ... (People) believe that higher impact factors must mean higher citation counts. But this is often *not* the case." [12]

Based on the above and various other facts, I agree strongly with the following assertion.

"India should create a new peer-review system, a new ranking of journals and new measures of impact – all tailor-made for our needs, problems, diseases, natural resources, and educational system. We need to believe in ourselves and not just chase world rankings – as individuals, as institutions and as a country. The enemy is within. So is the solution." – Prof. R.Gadagkar, IISc. [13].

The next chapter explores some ideas related to effecting the needed changes. Let me caution the reader, however, that not only based on my firm belief that no one has a monopoly on wisdom, but even more importantly on the fact that I may not be aware of many of the limitations or current efforts in India, these are offered more in the spirit of items for consideration and not as definite prescriptions.

## SUMMARY & ACTIONABLE ITEMS

This partial list is based on the discussion in Chapter 4. Please read these in conjunction with the recommendations in Chapter 5.

**Good Citizenship:** Develop and implement a standardized, secular curriculum enhancing morals, ethics, democratic values, citizenship responsibilities, and the rule of law. Any plan, system, or rule can succeed only if people at all levels exhibit a high level of probity, and national consciousness, cohesion, and purpose.

**Goals & Metrics:** Develop a top-down approach to education starting from the needs of the nation. Do not be satisfied by some gross numbers like literacy rates, General Enrollment Ratios, number of PhDs, and the like. Do not chase rankings in some international list without vetting the relevance and metrics governing them. The goals of Indian education should reflect specific needs of India in terms of the required labor force and its quality, higher international competitiveness, greater levels and quality of innovation and research, new business formation, and increased national prosperity. Develop metrics closely aligned to the goals and develop systems to track them continuously. Turn reporting from an annual ritual to an online tool for continuous improvement.

**National Cohesion:** Develop a national plan with cohesion across all states and union territories so that regional disparities can be reduced. While states may still retain much control, the Center needs to be more involved in making comparative assessments and in controlling grants strictly in ways to enhance higher performance.

**Bureaucracy:** Increase involvement of specialists in education and education management divesting control away from generalists and bureaucrats. Increase the pool of such specialists who can manage for rapid prototyping and improvement.

**Merit Orientation:** Infuse a culture of merit in admissions and appointments. Increase transparency and publish comparisons across institutions, districts, and states to reduce corruption and abuse.

**Evaluations:** At all levels, develop evaluation schemes to track progress continuously in terms of metrics closely aligned to the goals. Design feedback loops to make improvements on an ongoing basis. Infuse greater levels of data driven evaluations and improvements supported by a scientific and quantitative approach.

**First Boys & Girls:** Implement schemes starting at early levels of education to identify the truly gifted and talented and to challenge and develop them fully.

**Help the Disadvantaged:** Show zero tolerance to bias impacting access by any qualified individual. Focus on leveling the playing field through the early identification of the promising and meritorious, but disadvantaged, candidates and offer to them special help by way of scholarships, training, and mentorship. Aim for uniform thresholds and remove entitlements that degrade overall quality and merit.

**Private Institutions:** Set minimum standards for facilities, staff, instruction, and results. Monitor and regulate private institutions to ensure quality and performance. Mandate reporting on faculty composition and student performance. Publish comparative statistical summary results including survey results for transparent and easy comparison of institutions to create a climate of competitiveness based on quality. Leverage the tax-exempt status to get compliance.

**Foreign Institutions:** Take steps to ensure that curricula and instruction meet good standards and compare favorably with parent institutions in the home country.

**Environment:** Effect greater involvement and influence of each local community in the education of its youngsters. Implement cost sharing at local community level, effective parent/teacher organizations, and oversight by a school board of citizens to increase availability of first hand information and intervention, and to move away from the culture of dependency on big government.

**Teach to Learn:** Change the current curricula and modes of instruction to allow greater exploration, experimentation, self-learning, and practice. At higher levels, create opportunities for internships and learning in a practical setting.

**Soft Skills:** Develop programs to improve the soft skills of students to make them ready for employment and the professions.

**Campus Discipline:** Do not allow politicization of campuses or unruly demonstrations of any kind while at the same time ensuring a high level of academic freedom to learn and critically examine all kinds of knowledge. Make sure universities are led by leaders deserving respect and that they are empowered fully to maintain discipline and decorum in their campuses without any political intervention.

## Notes and References

[1]  Jean Drèze & Amartya Sen: *An Uncertain Glory, India and Its Con-tradictions,* Princeton University Press, Princeton & Oxford, 2013.

[2]  Shail Kumar: *Building Golden India – How to Unleash India's Vast Potential and Transform Its Higher Education System. Now.,* ONS Group Press, Fremont, 2015.  ISBN: 978-0-9966168-0-5.

[3]  Donald Kennedy: *Academic Duty,* Harvard University Press, 1977.

[4]  Thomas J. Stanley & William D. Danko: *The Millionaire Next Door: The Surprising Secrets of America's Wealthy,* Pocket Books, a Division of Simon & Schuster, 1996.

[5]  Robert T. Kiyosaki: *Rich Dad Poor Dad: What the Rich Teach Their Kids About Money That the Poor and Middle Class Do Not,* Warner Books, NY, 2011.

[6]  *Educational Statistics at a Glance,* Government of India, Ministry of Human Resource Development, Bureau of Planning, Monitoring & Statistics, New Delhi, 2014.
http://mhrd.gov.in/sites/upload_files/mhrd/files/statistics/EAG2014.pdf

[7]  "Tamil Nadu: Three Suicides over Fees but No Rooms, Teachers in Medical College, *The Indian Express,* Jan 25, 2016.
http://indianexpress.com/article/india/india-news-india/tamil-nadu-three-suicides-over-fees-but-no-rooms-teachers-in-medical-college/

[8]  "Chennai Engineering Grads Rank Low on English," *The Times of India,* Aug 14, 2015.
http://timesofindia.indiatimes.com/home/education/news/Chennai-engineering-grads-rank-low-on-English/articleshow/48474929.cms?messageid=39879998&intenttarget=no&r=1439497168374

[9]  "TN Finds Race to IIT Admissions Tough,", *The Hindu,* Nov 18, 2015.
http://www.thehindu.com/news/cities/chennai/tn-finds-race-to-iit-admission-tough/article7890080.ece?ref=tpnews

[10] P.J. Hilton: "Teaching and Research: A False Dichotomy," *The Mathematical Intelligencer,* Vol. 1, Issue 2, 1978.
http://link.springer.com/article/10.1007%2FBF03023064#page-1

[11] "Kakodkar Committee Fixes Target of 10,000 Ph.D. Scholars Per Year," *The Hindu*, Mar 16 2013.
http://www.thehindu.com/todays-paper/tp-national/kakodkar-committee-fixes-target-of-10000-phd-scholars-a-year/article4514553.ece

[12] "Citation Statistics," A Report from the International Mathematical Union (IMU) in Cooperation with the International Council of Industrial and Applied Mathematics (ICIAM) and the Institute of Mathematical Statistics (IMS), 2008.
http://mfa-national.org/yahoo_site_admin/assets/docs/Documents/CitationStatistics-FINAL-1.pdf

[13] "Research Management: Priorities for Science in India," *Nature - News & Comment, May 2015.*
http://www.nature.com/news/research-management-priorities-for-science-in-india-1.17509

## Chapter 5

# Improving India's Educational Systems

Education helps to bring out the best in human beings and lets society maximize benefits from its human capital. It prepares one for participation in society and the economy as a productive and contributing citizen. It elevates one's thoughts and vision. It empowers one to exercise one's rights and to gain one's dues from various systems. Thus, basic education (through high school) must be recognized as a fundamental human right. Refer to Drèze and Sen [1] for an excellent catalogue of the many benefits of basic education, both to individuals and to societies.

Although basic education should be viewed as a fundamental right, due to limited resources and the need to maintain high quality in what is produced collectively, societies have to consider higher education (college/university) as a privilege to be earned. They have to impose certain access restrictions based on aptitude and academic performance. Much of the challenge in Indian education today is in achieving a right balance between wide accessibility (fairness) and quality. An ideal system attempts to maximize fairness and accessibility while maintaining a high threshold of quality. India has attempted at fairness through a scheme of reservations and quotas at all levels based on criteria like caste and religion and has also lowered standards for selected subgroups to fill reserved seats.

My discussion below is conditioned by the following beliefs.

(a) Every human being should be entitled to education that empowers him or her to achieve his or her full potential and to contribute, to the best of his or her aptitude and ability, to society and to derive from it his or her fair (not necessarily equal) share.

(b) Education beyond a mandated minimum level will necessarily be discretionary and of limited availability, primarily due to lack of resources. Therefore, access to higher education cannot be universal. It is, however, essential that access is governed by fair academic criteria and a high level of transparency.

(c) The criteria for selection for higher education should be merit based and quantified by metrics that correlate highly with the success potential of the entrant not only in the program but also with regard to later contributions.   No one who meets applicable thresholds should be deprived of an opportunity for other reasons like economic status or membership in a particular subgroup.

(d) The system should not create or perpetuate any systemic disadvantages for subgroups (for example, those defined by a specific religion or caste) in gaining access, including at higher levels.

(e) Society should actively remove all systematic handicaps suffered by its subsections. But, those efforts should be directed to leveling the playing field by identifying the potentially meritorious among the disadvantaged groups early on and by providing them extra help (scholarships, training, and mentoring) as needed to compete as equals with others. Specifically, there should be no reservations and quotas, or lowering of standards to fill quotas. To borrow a phrase from Sen [2], quotas and reservations may only "have smallness thrust upon the young" by forcing systems to aim lower to accommodate the less qualified.

(f) A democracy should give free choice to its people. Therefore, private educational enterprises should be allowed in education. But they should be regulated to ensure that they maintain high quality and do not vitiate fairness by limiting access only to those who can afford an exorbitantly high price. Thus, it behooves the government to mandate that each private educational institution accept a certain percentage of poor entrants who meet the merit thresholds but are unable to pay the fees and bear other expenses imposed by the institution. The funding of such students should be a shared responsibility of the state and the private institution.

(g) The criteria for identifying disadvantaged groups should be primarily economic, although consideration may be given to obtain, in various professions and in government, a fair representation of subgroups not adequately represented. A functioning democracy certainly needs to be sufficiently representative of its people. As mentioned earlier, the attempt to achieve representation should, however, not be through reservations and quotas, but through giving out assistance in the form of special training and mentoring to promising candidates identified sufficiently early along with rewards for those spearheading such efforts.

India currently has a system of reservations, quotas, and lowering of thresholds for select subgroups. The motivation for adopting such a system soon after India's independence may have been noble or necessitated by the divisive politics of the erstwhile colonial ruler who was a master of the art of "Divide and Rule." But unassailable evidence exists to show that it has created serious problems of various types including an erosion of standards and quality and considerable brain drain out of the country. Recently, it is also causing great social unrest among those suffering a reverse discrimination. A change must occur soon. However, since any sudden change would bring in violent reactions, a firm policy should be evolved to move to a merit based system as envisioned above within a few years (say ten) with planned periodic reductions of the levels of reservations and quotas, and simultaneous introduction of more constructive remedial measures of support based on less divisive criteria than caste and religion. This is not a small challenge in today's India where national interest is highly compromised for parochial and party interests.

Having laid down the basic beliefs underlying my recommendations, I am now ready to get to some specifics.

## 5.1 Basic Education

Not only in developing nations like India, but all over the world, educational systems are suboptimal to varying degrees. Systems have not kept up with sociological and technological changes in a timely manner. Most adopt a "one shoe fits all" approach that results in an inability to challenge highly talented students adequately. They also exasperate others with unrealistic challenges to even points of failing or dropping out altogether. With a mistaken notion of equality and fairness, they try to aim for a middle ground, but in reality serve no group to the best of what could be achieved. Many highly talented students are forced to regress to the average. Worse still, large numbers of students who graduate out of middle schools (and even high schools) are functionally illiterate and are, therefore, neither employable nor capable of pursuing higher studies.

Systems need to recognize that despite all efforts of parents, educators, and educational institutions, not all students will evolve in an equal and uniform manner in ability, aptitude, or performance. Yet, society should educate everyone so that he or she can have the opportunity to realize his or her fullest potential later on in life. That certainly requires some customization of education and streaming of students into different tracks based on objective criteria. School curricula and organized education, in general, do not seem to reflect this need adequately.

As an example, I would refer the reader to the content of Common Core Algebra I (a required course) as seen in the examination [3] of January 2016 for high school students in New York. (I pick an example from a society that spends an enormous amount of its resources on education to illustrate that even in such societies the situation is less than optimal.) It asks students, among others, to do the following: identify which of a set of given expressions would result in an irrational number; identify a specific quadratic equation that would increase the area of a 10x12 rectangle by 50% with equal increases of the lengths of its sides; factor a cubic polynomial; and determine which functional relationship looks linear based on a set of residual plots resulting from a linear statistical fit to the data. Despite being a mathematician and my love for mathematics, I cannot help asking how important this knowledge is for a whole host of students who will not be pursuing college level studies in science or engineering? Would someone who would like to be and could very well be a great poet or journalist, or even a good car mechanic, electrician, or plumber running his own shop some later day (and, hopefully, even getting rich thereby) really need this type of knowledge? Could such students be expected, with any reasonableness, to show the aptitude to learn such material, or would they be just set up to drop out or fail? Every high school student in New York has to take Algebra I or Geometry and get a 65% passing score. Geometry, of course, has its own formalism, axioms, theorems, and corollaries. Most readers may find the questions in it even more daunting than the ones from algebra I cited. I am sure my reader in India or in any developing nation will find even more bizarre parallels than the ones I have cited, for systems there offer even less of a choice to students than in the USA with respect to courses they can take in their schools.

As noted in [4], there is considerable concern among educators to "ensure that the curriculum does not lead to a fragmentation of the student body." Thus, there is an attempt to force the students to take the same set of courses, or in other words, to make the curriculum very centripetal in character so as to draw students together. Unfortunately, that turns out to be counter-productive and goes against nature, as people are different in a variety of ways. The mindset should be one of letting children grow differently but with mutual respect and appreciation for each other's domains of interest and strength.

Thus, I believe that a good place to start is with a top down approach as to what the nation wants to achieve through educating its people both at the collective and at the individual levels. Uniformity cannot be a goal and should not be pursued mindlessly. There is no denying that there is a minimum level of knowledge that every human being requires to function and thrive in society and to carry out various day-to-day transactions effectively. Examples of these include the ability to write various types of letters, fill out an application form on paper or on the Internet, understand road signs and rule books such as those related to driving, conduct some rudimentary business and business transactions, grasp some basic knowledge of the role of money and credit, balance a check book, understand profit and loss, understand some basic economic notions such as principal and interest, read a newspaper or magazine article with reasonable comprehension, operate a personal computer or tablet to interface digitally with the world to research and gather information as well as to communicate. In addition, everyone must have knowledge of basic numeracy and some rudiments of government and the law. Today, even a street vendor and a store clerk are forced to acquire many such skills to stay ahead of the curve. A well-designed educational system should provide this level of minimum functional knowledge and skills to everyone by the time they finish middle school (the eighth grade).

Such education should incorporate sufficient hands on practice components, means to develop ability for self-exploration of knowledge, and some challenges to think out of the box and to innovate. The goal, at the early levels of education, should be not a stamp of pass or fail, or ranking against peers, but a true identification of the right groove that one fits well and in which one can grow to be a success with happiness and a sense of fulfillment. In addition to the core that teaches basic life skills of the type noted above, there should be a slate of optional courses for those who are able to learn and demonstrate mastery in the core curriculum at a faster pace. The latter is the customization part.

Evaluation at these early stages should be frequent and should provide an assessment to parents and to the school of the strengths and weaknesses of individual students. Periodic assessments should also identify subsets of curricula and instruction needing improvement as well as teachers in need of additional training.

It is not clear what fraction of those passing the eighth grade in India really master the minimum I have cited above or manage to find what really tickles them. Part of the reason appears to be that school curricula in India are biased by what educators would call "Academic Rationalism" [4] which makes basic education strongly subject centered and not functionality centered. They emphasize the disciplines' concepts and syntax rather than "Personal Relevance" [4] centered on student needs "to enable students to find personal meaning in what they study." The majority of systems also fail in the area of a needed "Cognitive Process Orientation" [4] that should be "concerned with helping students solve problems, develop better thinking skills, and learning how to learn."

Although it is beyond the scope of this book at its popular level to get deeper into theories of school curricula, the above discussion can be seen as clearly establishing the need for a fresh look at primary and middle school curricula in India. The first step is to identify what exactly are to be achieved at various levels and then to determine the specifics of the curricula. Through some core and optional courses, it must equip all students to acquire the minimum needed life skills, to determine their own interests and skill levels, and then to move on to an appropriate high school stream suited to their aptitudes and abilities.

Primary education has to be necessarily more structured and uniform. But, even at the middle school level (Grades 6-8), the curriculum should begin to incorporate a bunch of optional courses beyond the core ones for those students who are able to master the core areas faster. This would necessitate courses to be modular with students progressing not necessarily as an entire cohort in step with each other as it occurs now on a yearly calendar. The optional higher-level courses should have a greater dose of "academic rationalism" and "cognitive process orientation" that will be appropriate for those students who will later pursue higher collegiate education.

Most importantly, based on objective criteria including psychological screening, the highly creative and precocious pupils should have, at least at a district level if not at the individual school level, special 'gifted and talented' programs that challenge them even more and help them to augment their innate creativity. These programs should, in essence, serve as a mild introduction to systematized exploration, experimentation, and creative expression. This type of a program is almost entirely missing in Indian school systems, and this results in a real failure to identify potential innovators and researchers early on. The consequences of the current tendency to concentrate on the middle is that the real crème of the crop is either abandoned or is forced to regress towards the average.

## 5.2 High School Education

At the high school level (Grades 9 and above), it is certainly important to channelize pupils into different streams. There should be separate tracks for science and technology, the arts and literature, other humanities, and one for vocational education. The assignment of students to these tracks should be based on aptitude, preference, and minimum thresholds of performance. While every track should get a sufficient share of what is taught in other tracks so as to come out well rounded with a good understanding and mutual respect for each others' areas, each track should show a high emphasis of its own focus in its core courses. Once again, within each track, courses should be separated into essential core courses that everyone should take, and optional ones that form meaningful subgroups. Thus, for example, in the science and technology track, certain core courses on mathematics, computing, physics, and chemistry could be mandatory for all while optional subjects could cover closely related areas like probability and statistics, modern physics, bio-chemistry and the like depending on the interest of the student. Those in the other tracks may get just one science course a year that would give them a grand tour of the different areas learned by the science and technology track of students, albeit at a more superficial level. The emphasis of such a tour would be on uses and relevance rather than on formalism and techniques.

With regard to vocational education, the first semester of high school could be organized just as a grand tour of the different areas one could specialize in. This could be followed by a few intense semesters in one or two closely related areas chosen by the student. These should either prepare the student for a job right after high school as a technician in an area or for pursuing a higher-level diploma or associate degree comprising of, at most, two years of additional study beyond high school.

Vocational education should not be aimed only at providing a labor force. With real imagination, it could be turned into one that forms the fertile field for future entrepreneurs, small businessmen, and innovators. Thus, in addition to specific vocational skills, students should be exposed to ideas of entrepreneurship and various business skills through both required and optional courses. A big missing component in the Indian economy is the skilled, business-savvy, individual entrepreneur in areas like electricity, plumbing and the like, who can provide high quality service to the public and earn a high standard of living. Given that it is this sector that really drives the economy to high gear and creates greater equality in wealth and quality of life, substantial importance should be placed on this even at the high school level.

## 5.3 Meeting the Challenge

The first challenge in implementing the above type of a top down approach is to ensure that children do not get locked into tracks too early. Instead, room should be given for reasonable flexibility to make changes. In other words, students in a given track should be allowed to switch to another track at least once by taking some optional courses and performing in them well. This could also address the reality that just as flowers do not bloom synchronously, children also do not find their right groove at the same age or in a given grade. The key to everything here appears to be flexibility, which is totally lacking in education, especially in India.

The second challenge, particularly relevant for India, is that, most often, parents choose the academic paths for their children. These decisions are not necessarily made based on children's aptitudes and abilities, or on the recommendations made by teachers and specialists, as they should be. Objective, transparent, and trustable assessments by competent professionals followed by parent counseling thus become extremely important. All that the schools can do is to lay down the criteria for entry into various programs, make them absolutely transparent, and enforce them equitably without fear or favor. At a higher level, the state can force schools to implement the processes fairly and efficiently through standardized testing. In addition, the state should publish results comparing schools and programs and force them to undergo periodic accreditation reviews so that parents can make informed and intelligent choices.

The third challenge is the crafting of a set of curricula, courses, and modules along with evaluation schemes. As noted in an earlier chapter, evaluations should be continuous and should include a feedback loop that would provide valuable information for improving curricula, instruction, and processes at individual and collective levels. The country should leverage much of the research within the country and outside and develop pilot programs and make them systematized and near perfect before rolling them out at a large scale. A nationally coordinated effort covering all states with high cohesion should be made so that development is more uniform across the country.

The fourth, and perhaps the most daunting challenge, is the creation of the army of teachers needed to implement such a visionary program successfully. It is here that it becomes important to raise the teaching profession to its deserved level of remuneration, reward, and prestige and to ensure that only the very highly motivated and qualified are hired and kept. The country needs to develop a greater culture of accountability and performance in this profession, and move away from permanent tenure, seniority based upward movements and the like that do not foster and reward good performance. Teacher recruiting, retention, and promotion should all occur based on criteria of merit and performance as evaluated by neutral evaluators both in and out of the classroom and should fold in student performance (with due filters applied to differences in groups taught by different teachers). Contrary to the unfounded fears among teachers, valid statistical procedures (such as propensity matching [5]) do exist that help to filter out the effects of other variables and help to evaluate teacher performance objectively. Despite being a nation with a top corps of statisticians, it is unfortunate that India is yet to take advantage of data driven improvements in the educational sphere. There is a great need to infuse more quantitative and scientific methods in the evaluation of educational systems and personnel at all levels.

The literature on effective teaching and teachers is extant. One of the excellent books providing a concise overview I have come across in the USA is that of Stronge [6] published by the Association of Supervision and Curriculum Development (ASCD). This is a vast area in itself. But there are some points that stand out clearly: (a) subject area knowledge alone is inadequate for a teacher, and a basic knowledge of pedagogy and pedagogical techniques is absolutely essential; (b) effective teachers exhibit a dual commitment of student learning and personal learning; (c) early mentorship in a classroom setting is the highest contributor to teacher training; (d) teachers typically take three years to gain mastery in their profession and beyond that period do not gain as much except through targeted professional development programs. The last point is particularly important as a guide to setting probationary periods for teachers. Again, change can occur only gradually, and there needs to be several pilots followed by gradual scaling up by subsections of schools based on current performance, geographic areas, and the like. Some scheme of starting with a highly trained team of "master teachers" whose members, in turn, train others is needed to create a geometric growth in the number of qualified teachers so that the process does not take too long to come to fruition.

As these changes occur, it is essential to infuse a greater level of educational management professionals in decision making at all levels, in schools, school systems, and government. That cadre should have enough representation from those who have hands on experience in education. That implies that high performing educators should have the ability to take a sabbatical from their work and move on to such higher spheres of influence. In short, the government should put in a concurrent effort to create a cadre of highly competent education professionals who can replace the current bureaucrats and generalists. The present bureaucrats will not be an adequate match to the real challenges, and the current state of affairs more than attests to that fact.

Like in many of the other areas we have discussed in the previous chapters, the amount of change needed is very large. But if we keep letting that deter us from even getting started along a right path, there is no hope of ever improving the current systems. With our goal to become a developed nation with advanced capabilities, it is absolutely essential to start the process soon even though we may not have the perfect systems right away or ever, and our ascent will be long and arduous. The above is but a broad-brush attempt at a new template, but as the saying goes, 'the devil is in the detail.' But, to further postpone action believing that the necessary leadership or solutions cannot be generated in a land of over a billion people who have collectively surmounted many daunting challenges would be a major folly. As a country, we have tried conventional wisdom in an unquestioning manner long enough. It is time to think afresh and adapt with our needs, our aspirations, and our ground realities in focus.

## 5.4 Improving Higher Education

Like school education, much of India's higher education is also in disrepair and needs a major overhaul. A major difficulty is that the big bulk, the legacy system of universities, which is in severe disrepair accounts for most (86%) of the students in higher education. Furthermore, their control is spread over the multitude of India's states with little central authority. But, even more importantly, the challenge for India is to make sure that the primary beneficiary of India's higher education is India - an issue that alone could give the *raison d'etre* for bringing in change but, nevertheless, gets hardly addressed.

The benefits that could be gained by India in developing its educational institutions are enormous. Just consider the following as an example of what a good educational system can contribute to an economy if only there is proper direction and resources given to it.

> "According to a 2012 study, since the 1930s Stanford entrepreneurs (faculty and alumni) have started 39,900 companies, which in turn have created 5.4 million jobs and generate US $2.7 trillion in revenues annually." – Shail Kumar [7]

Note that the revenues mentioned above are more than a third of India's estimated GDP for 2014. The companies above include some great and well-known ones like Hewlett-Packard, Varian, Cisco, and

Google. What is worth noting is that many of the start-ups that have grown to be giants in the USA were indeed built by Indian immigrants. Also, many of the industrial giants of America have gained and continue to gain much from the contributions from NRIs and their children. What a difference would it make for India if only we could make such great things happen right in our own country and if Indians in India were to be the owners of such great enterprises? India should not waste any more time in working towards that lofty goal.

My task of enumerating some specific changes needed in India's higher education has been simplified somewhat by the recent appearance of the book by Shail Kumar [7]. The discussions there of how some leading universities in the USA transformed themselves into their exalted positions have many lessons for India. Kumar identifies many best practices India could attempt to emulate in the area of higher education and lays down a list of prescriptive recommendations. The material here may be viewed as augmenting some of those discussions, although, even at a basic level, I do differ from Kumar with regard to my approach.

My approach is based on a planned, gradual reform over a reasonable period of time (like a decade) combined with a few immediate reforms that will bring short-term returns. My approach also attempts to leverage existing assets as much as possible and does not base itself on the assumption that India can afford a very large budget increase such as, for example, to create a number of new mega universities. Even if the money could be found, finding the necessary human resources and dealing with the rehabilitation of a large army of faculty and systems currently mired in inefficiency can be significantly challenging. And, even more importantly, there is the *a priori* need for making a strong case for investing for change based only on

India's potential gains. The first place to start, therefore, is by ensuring that India gains fully from whatever effort it puts in so that the effort itself becomes justifiable amidst the competing priorities facing the nation.

## Poor Returns

The Indian taxpayer subsidizes India's higher education to an incredible extent. Consider the Indian Institute of Science (IISc), which is an educational institution of international standing that imparts high quality of education and accounts for training many scientists, engineers and academics of Indian origin now in the Western hemisphere. Even a cursory glance at its 2016-2017 fee schedule [8] should leave an American totally dumbfounded. According to [8], the fees for a science degree program is Rs. 40,000 (about $597) for three years combined; Master of Engineering Rs. 27,400 ($408) for two years combined; for a Master of Management Studies program Rs 302,000 ($4507) for two years combined. And, many students do get scholarships or interest-free loans. As for the Indian Institute of Management, Ahmedabad [9], from where large American financial corporations handpick graduates each year, "No tuition fees is charged for Indian students. Selected candidates will get a stipend of Rs. 29,400 [$438]per month..." Compare this to the $60,000 or more of undergraduate fees per year even in some moderately good universities in the USA, and the $100,454 yearly fees for MBA at the Wharton School [10].

Unfortunately, with the crème of the crop of India's graduates migrating abroad or working for foreign multinational corporations, the direct returns to India are very low from India's investment in higher education. While there may be some indirect returns to the nation by way of international recognition of Indians as a talented people with scientific and technological merit, those are not sufficient. Nor has India been able to get significant support and help from the majority of Indians who have left her shores and have done well. Part of the blame does rest with India's educational institutions, which have not done a good job of tracking their alumni and finding ways to keep them plugged in through various alumni related programs and awards. This is an area where India has much to learn from US universities.

India needs to take immediate steps to ensure that its investment in the higher education of each Indian does pay significant returns to India and simply does not end up enriching other nations at its expense. Certainly, India needs battalions of "first boys and first girls," but not to be given away as adoption babies to others. While any reasonable estimate of the total wealth created for other countries by those of Indian origin would be astronomical, even the wealth being created for them in India by Indian trained engineers and scientists is of such a high magnitude that India needs to evolve a strategy to redirect all that potential to its own progress and development. A beginning must be made immediately.

It is important that India develops a cost accounting for higher education by at least institution categories. It must charge a high fee for higher education. That fee should be commensurate with those in countries that hire most of its graduates, or at least high enough to cover per student costs with a pre-defined cover for inflation. Students should be given a choice to pay that amount each year or to take it out as an interest-free loan from the government. For each year worked by the graduate as a self-employed person in India paying income tax above a certain threshold, or for a majority Indian owned enterprise, or for government, the person may be given a waiver for a part of the loan (say, 10%). If a person fails that test, then the entire amount of one's fees should become payable with interest in installments within a stipulated period (ten years at the most). It is fair that those who seek fortune without benefiting India directly, and are enabled to do so by Indian education, pay back a fair share to India.

It is not enough to lament that we lose our talented people to other nations. Honestly, if the many who have left had stayed back, the unemployment situation and unrest in India would be much worse due to the lack of opportunities and, even more importantly, due to the unfair and unconscionable ways in which the limited pie is apportioned, with merit playing a minimal role. There is a strong case for India to first create a climate where its top graduates can and will be absorbed by the nation into employment that makes it less attractive for them to flee for acquiring personal wealth and good quality of life elsewhere. Even more than money, the corruption, bureaucracy, and other major irritants need to be redressed to retain even those true scholars whose primary lure is not wealth. Just increasing the number of graduates without addressing these issues may not bring back rewards to India, although it will certainly benefit many Indians as individuals and also other nations. Herein is my personal skepticism for calls such as those in [11], [12] for increasing the number of Indian PhDs by an order of magnitude.

If the government is not capable of tracking and enforcing liabilities incurred by individuals for higher education, a commercial entity like a bank should be brought in as a middleman for a fee to enforce the contracts. Some special schemes like loan forgiveness for graduates who engage in notable and verifiable social service, make major scientific discoveries that benefit India (not necessarily in monetary terms), create successful commercial innovations, or create jobs for others by starting businesses should also be developed to serve as additional incentives for graduates to go beyond traditional employment. These will draw more Indian graduates to work for India's direct benefit and on increasing Indian innovations. These schemes will also result in preventing foreign MNCs from enjoying an unintended, but substantial, subsidy from India. Most importantly, it will enhance the chance that the research and inventions of India's graduates enrich India and not some other country. India needs to protect and leverage its intellectual assets, particularly human capital, at least as vigorously as it would its material assets, and, preferably, even more since they are the key to derive the best out of other types of assets.

Every rupee spent by India on graduating a student through college is at the expense of a rupee that could help to eliminate a sharecropper's suicide, or the preventable death of a child or of a new mother. It simply cannot be given away as largesse even to India's "first boys and girls" especially if India does not stand to benefit from them. Anyone who has benefited from such largesse has to become more sensitive to this reality. All the arguments given earlier apply equally to those in private educational institutions in India since those institutions do gain much through their non-profit status, and through the supporting infrastructure and oversight paid for by the taxpayer. Basic education is a fundamental right and may, therefore, obligate a state to subsidize it heavily, but higher education is a privilege for which one should be ready to pay a fair price. The state should certainly ensure, however, that no qualified person is deprived of access to even the best of the systems due to one's economic status. The current system of near free higher education with no expected payback from its graduates cannot help India become a developed nation. It only serves to widen the gap between India and the countries benefitting from its highly talented graduates.

### Increasing the Availability

I do not believe that India can quickly build a number of mega universities, and even if it did, still many current institutions that are lagging in quality will have to be dealt with.  In the short run, the approach should be based on identifying the best resources that are available and creating a sharing process.  That would facilitate an up-lift of the laggards and, thereby, result in an increase in the quantity of quality education available to students.  Some specific efforts yielding results quickly should begin to be pursued immediately even as plans are made for much larger efforts over a longer haul.  Let us examine a few of them.

The elite institutions of India like the IIMs and the IITs are national resources.  They should be involved in developing faculty at other institutions to bring the latter on par and, thereby, become change agents.  Opportunities should be given to selected faculty from various institutions to attend some classes in these elite institutions as guests so that they may improve their own repertoire and teaching methods. Faculty on both sides taking such initiatives should be rewarded based on quantified metrics of impact.  In addition, the elite institutions should throw open their seminars and guest lectures, at least through live multimedia and recorded programs, for other institutions and their faculty. Teaching loads in colleges should be reduced to enable faculty to participate in professional improvement activities of this type.  That holds for library and laboratory facilities as well. Territorial attitudes should not be tolerated from any quarter.  [As a young lecturer in Loyola College,  I remember that when I made a request to the director of an elite institute in Chennai to let us know of upcoming lectures by visiting foreign scholars, I got rebuked with the comment, "Sir, the mountain does not ever go to the Mohammed"].  Just like in the case of the training of school teachers, what is needed is the creation of successive cohorts of trainers of university faculty who will transform the vast majority of university faculty to deliver quality comparable to their peers in the elite institutes. Needless to say, many faculty members in universities are already on par with those in the elite institutes. They should be actively identified and made partners in the endeavor to increase the pool of trainers.

No selection process is perfect, and there are days when even the best performs poorly. Many students who are equally bright or even brighter than the lucky ones who get admitted into the elite institutions can get stuck in less stellar places for a variety of reasons. [I have the dubious distinction of failing to get into ISI; the type of entrance test based on multiple choices was totally unfamiliar to me when I took it. Not that I should complain, for it forced me to learn much on my own and enabled me to become a stronger researcher and go much farther than I probably would have, had I grown in a more structured environment.] Since mass level screening processes have to be employed and may still be imperfect, ways should be found for the very top students in the non-elite educational institutions to take at least some select courses from the elite institutes. The overheads involved here are minimal, especially in the case of classroom participation without laboratory work, and even less if multimedia technology can be brought in. Alternately, the elite institutes could select and offer a set of courses for top students from other institutions. These students could be chosen based on objective criteria and interviews.

Given the disconnect in control that exists today with the elite institutes under central control and the universities under states' control, some new mechanisms or rules will have to be devised to ensure that the universities cooperate. An approach based on both carrots and sticks should be evolved to ensure that cooperation is indeed obtained among the various actors, since what is at stake is India's ability to break the fetters that hold her back from becoming a developed nation.

## Curricula and Courses

A concerted effort should be made to break the silo mentality in curricula, courses, degrees, and faculty. Through inter-departmental programs and optional courses, the menu available to students should be widened to allow greater cross-pollination in training and idea generation.

Teaching should involve more hands-on practice, a significant orientation towards applications, opportunities for exploratory learning, and engagement with unstructured problems. An effort should be made to get students engaged in industry internships to complement their academic training. Certain new formats like industry co-op programs such as the one successfully implemented at many universities in the USA like Drexel University, Philadelphia, wherein students alternate between the university and industry in alternative semesters, should be explored.

Engineering societies should actively promote India-specific professional certifications that can help to bring in uniform levels of competence and learning across the board in all geographies and institutions. The goal of engineering education should not be viewed as getting people ready for employment, but should train, at least the motivated and the best, to become entrepreneurs and businessmen. That requires the availability of optional courses imparting various business and management skills.

## Faculty Development

Positions in higher educational institutions should be highly competitive, and appointments to them based on demonstrated performance and future potential as educators and researchers. Market-based salaries should be paid based on the field of specialization and the level of attainment so that the best people are attracted to the positions. A reasonable level of de-coupling of salary levels and ranks should be effected so that people shine with what is their forte and do not let the infamous Peter Principle [13] force them into positions of incompetence. This happens, for example, when some good researchers get promoted to positions with considerable administrative responsibilities for which they are not suitable. A meaningful portion of the compensation should be made discretionary (similar to bonuses for company executives) and given out strictly based on demonstrated excellence and contributions. Also, a university or teaching institution cannot become the last resort of the least employable. Even in situations where accommodation is made for special purposes, for example, for new PhDs returning from abroad, a probationary period (say, of three to five years) has to be set aside, within which time the appointee has to show tangible and verifiable performance.

All types of entitlements like seniority-based promotions need to be abolished to instill a culture of performance and performance based rewards. Incentives and rewards should be given to those who go beyond the call of duty and contribute in significant ways. Within minimal bounds to ensure that one's performance in one's primary capacity does not degrade, and in the absence of demonstrable conflicts of interest, faculty members should be encouraged to involve in activities like consulting for industry and to be compensated for such engagements. Similarly, institutions of higher learning should foster a climate of intellectual property generation and licensing with shared monetary rewards. If anything, that would only improve the application focus of faculty and help to bring more practical aspects into teaching and research.

## Practical Orientation

A concerted effort across the board in science and engineering must be made to increase the application focus of curricula and of faculty members. The lack of hands-on experience in industry is a major limiting factor for Indian faculty engaged in higher education. That needs to be remedied in a variety of ways. One of them is a paid sabbatical for faculty to spend a semester or a year in an approved industry based on a proposal that demonstrates some specific potential benefits. Educational institutions need to bring in industry experts, not only to interact with their science and engineering departments, but also to design and teach some specific courses with greater applications content of practical relevance. Where possible, NRI faculty going on sabbaticals and recent local and NRI retirees should be sought after for augmenting the faculty, with preference given not to badges and medals but to service orientation, and competence in both theory and applications as evidenced by recent publications, patents, and industry experience.

Training, especially in laboratory and practical methods, should be organized for university faculty on a regular basis. Those graduating from such programs may be incented through salary increases and various visible recognitions so that these programs are taken seriously and do result in improvement in university education. Certain minimum continuing professional development should be enforced on all faculty members so that in this fast changing world of technological and scientific development, one does not become a fossil living on past glory alone. (This is a serious challenge in US academia where systems cannot even mandate a retirement age.) The requirement of professional development should be flexible so that it could be met not necessarily by taking courses, but also through demonstrable expertise (e.g., new courses designed and taught, consulting, patents, or publications) in current areas of research.

## Research Quality

A targeted effort must be made to improve the quality of research training and the quality of dissertations. Needless to say, recruiting only highly qualified scholars is an essential first step. Research is not a place where mediocrity or entitlement can be tolerated because even a semblance of that has the infectious effect of bringing about large-scale demoralization. As for dissertations, in the majority of instances, the choice and formulation of the right topic for research and the hypotheses and theories to explore are the ones that determine the quality of the final output. [Inadequate attention paid to this has landed me, more than once in my capacity as external examiner of some doctoral dissertations from India, in the unenviable situation of inflicting pain on a student for the lapses of a mentor]. The exercise of proposal writing should also be an attempt to help the student to identify an area in which he or she can generate much further research beyond the doctoral dissertation, and not just a specific problem of narrow scope. A PhD should not be viewed as a terminal exercise, but as the potential beginning of a research career lasting a lifetime. Therefore, enormous attention should be paid to increasing the quality of research efforts at the proposal stage so that the scholar starts off on the right foot. Having a committee with members from related departments, and even outside, such as from industry, who will ask incisive questions and suggest additional ideas, is absolutely necessary to avoid a generational decline in the quality of research. Such decline is inevitable if dependence is mainly on one's mentor for topics and guidance. A PhD thesis need not be earth-shaking research *per se*, but it certainly must have enough novelty to differentiate itself from a (challenging) class project and establish the candidate as a fine thinker.

Funding agencies should identify areas of importance, and within them subtopics, and invite proposals from faculty in a way that the funding process drives research towards topics of value and high quality. Herein also lies the opportunity for the nation to bring the talent resident in the country to solving India's problems.

Certainly, if only for maintaining and increasing the credibility of India in science and technology, India needs basic research and theory of the 'pure' kind, and many who will shine in them to win international level accolades. Besides the irrefutable argument, "Man does not live on bread alone," newer theories often give birth to new branches of knowledge and result in newer applications too. But there is also a parallel need for India to solve its own problems. Thus, a right balance must be struck between international visibility and national needs. To state it without sugar coating, we need to eliminate much of Indian research that fits into neither of these bins. There is a need to steer the efforts with a greater sense of purpose to result in the happy situation that enough applied research gets done to increase our ability to support basic research in addition to solving many national problems.

## SUMMARY & ACTIONABLE ITEMS

### SCHOOL EDUCATION

**National Cohesion:** Develop a high level of national cohesion on all plans of improvement and implementation so that regional disparities are minimized. Plan for a phased implementation scheme starting with small pilots and selected schools with high potential of success, so that within a stipulated period (say ten years) education across the nation will be transformed. Publicize early successes to increase awareness and motivate local communities to move forward.

**Community Participation:** Involve local communities through Parent/Teacher Associations, Local School Boards comprising of citizen volunteers, and discretionary funds at the disposal of such boards to improve oversight, local control, and information flow. Assess community satisfaction with institutions through periodic surveys and publicize the results for easy comparison. Develop appropriate cost sharing plans so that local communities have greater stake and a say in the management of quality in their schools.

**Curriculum:** Re-design the primary and secondary education curricula emphasizing functional skills instead of being centered on subjects and their concepts and syntax. Increase participatory learning by students and reduce the level of unidirectional instruction from teacher to students. The goal should be to ensure that by the end of the eighth grade, students have mastered essential life skills and are prepared for high school.

**Modularization:** Modularize content so that pupils can move at their pace. Specifically, allow fast and better students to move faster and to acquire additional knowledge and skills while allowing a little more time for those lagging behind.

**Optional Topics:** Starting middle school (the sixth grade), introduce optional modules or courses for the benefit of the more accomplished students.

**Gifted & Talented Programs:** Identify and select the unusually gift-ed and talented children and place them in special programs challeng-ing them to the fullest and complementing their innate talent with systematized means of exploration, creativity, and expression.

**Streaming:** Develop a slate of different tracks for high schools (Grades 9 and above) and stream students into them based on performance, ability, and aptitude. Make room for reasonable flexibility so that students do not get locked into tracks prematurely and and are able to switch, if necessary.

**Entrepreneurship:** Do not view vocational and terminal tracks in high school as geared only towards generating a paid labor force of future employees. Develop optional courses on entrepreneurship and business skills to students in vocational tracks who may not proceed beyond high school so that they have a chance to grow into being skilled owners of small businesses providing quality products and ser-vices to society.

**Teachers:** Train a set of master teachers who, in turn, can train others in the new curricula and methods so that the competent teacher pool grows geometrically covering the entire nation.

**Management:** Wrest control and decision-making away from gener-alists and bureaucrats and increase the involvement of educational spe-cialists and those with hands-on experience as educators. Increase the pool of qualified personnel to take on the education management re-sponsibilities at all levels. Create innovative programs like sabbaticals for exemplary teachers to grow into effective administrators with greater influence.

## HIGHER EDUCATION

**Incent Indian Graduates to Serve India:** Create a climate where top Indian graduates will have less of a motivation to flee India or

Indian employers by improving the overall quality of life and eliminating irritants like corruption and the unfair ways in which the limited opportunities are given out without regard to merit as the primary criterion.

**Make Higher Education benefit India:** Nominally charge each student appropriate fees based on actual cost and comparable costs in other countries, but treat it as a loan. For each year of involvement in the economy in a manner benefiting India directly, such as being employed in government or in a majority Indian owned enterprise or creating employment as an entrepreneur, a certain portion of the loan may be forgiven. But otherwise, the loan should become repayable over a few years with interest.

**Alumni Relations**: Improve alumni relations by institutions so as to create a greater bonding by NRIs to India and their parent institutions. Find creative ways to tap into NRIs as a resource.

**Share Resources:** Share quality resources in elite institutions with others so as to uplift the latter. These include faculty, library, seminars, and even classes. Allow top students from other institutions access to lectures and classes in elite institutions.

**Faculty Improvement:** Make appointments merit based. Bring salaries up to reflect market salaries by specialty. Uplift lagging faculty by allowing them as guests in classes of superior faculty members in elite institutions. Create a succession of trainer cohorts who will train faculty to improve their instruction in diverse ways. Incent participants with rewards. Remove entitlements like promotions and salary increments based on seniority. Consider making part of the emoluments as bonuses to be earned through quality performance. Overlap salary ranges across levels so that one is not forced to move up to a job level one is not suited to just for higher salary. Involve people with industry experience, demonstrated practical orientation, and innovation success.

**Curriculum:** Break the silo mentality in curricula, courses, degrees, and faculty. Infuse greater applied and practical content. Focus also on self- learning, exploratory studies, and ability to consider ill formulated problems. Encourage professional certifications by a common entity as a means of enforcing minimal quality and content across all institutions.

**Research:** Improve research quality by focusing early at the proposal stage and by involving a committee drawn from several areas and, where appropriate, from industry. Use funding to direct research into high impact areas and into areas of national importance and urgency.

## Notes and References

[1]   Jean Drèze & Amartya Sen: *An Uncertain Glory, India and Its Contradictions,* Princeton University Press, Princeton & Oxford, 2013.

[2]   Amartya Sen: "The Importance of Basic Education," *The Guardian,* Oct 28, 2003. http://www.theguardian.com/education/2003/oct/28/schools.uk4

[3]   "The University of the State of New York, Regents High School Examination, Algebra I (Common Core)," 2016. http://nysedregents.org/algebraone/116/algone12016-exam.pdf

[4]   Allan Glathorn & Jerry Jailall: "Curriculum for the New Millennium," Chapter 5 in *Education in a New Era,* Ronald S. Brandt (Ed.), Association for Supervision and Curriculum Development, Alexandria, VA, USA, 2000.  ISBN: 0-87120-363-4.

[5]   Peter C. Austin: "An Introduction to Propensity Score Methods for Reducing the Effects of Confounding in Observational Studies," *Multivariate Behavioral Research,* 46(3): 399-424, 2011.

[6    James H. Stronge: *Qualities of Effective Teachers,* Association for Supervision and Curriculum Development, Alexandria, VA, USA, 2002.  ISBN: 0-87120-663-3.

[7]   Shail Kumar: *Building Golden India – How to Unleash India's Vast Potential and Transform Its Higher Education System. Now.,* ONS Group Press, Fremont, 2015.  ISBN: 978-0-9966168-0-5.

[8]   "Indian Institute of Science [IISc] Bangalore, Karnataka, Fees Details 2016-2017." http://collegedunia.com/university/25603-indian-institute-of-science-iisc-bangalore/fees

[9]   "Indian Institute of Management, Ahmedabad, Fees and Expenses," http://www.iimahd.ernet.in/programmes/fpm/how-to-apply/fees-and-expenses.html

[10]  "Wharton, University of Pennsylvania – MBA Program, Fees," https://mba.wharton.upenn.edu/financing-mba/tuition/

[11]  "Send 10K PhD students each year to US: Narayana Murthy," *The Times of India,* Jan 30, 2016._http://timesofindia.indiatimes.com/

business/india-business/Send-10k-PhD-students-each-year-to-US-Naray-
ana-Murthy/articleshow/50780069.cms

[12] "Kakodkar Committee Fixes Target of 10,000 Ph.D. Scholars Per *Year,"*
*The Hindu, Mar 16 2013.*
http://www.thehindu.com/todays-paper/tp-national/kakodkar-com-
mittee-fixes-target-of-10000-phd-scholars-a-year/article4514553.ece

[13] Laurence Peter & Raymond Hull: *The Peter Principle: Why Things
Always Go Wrong,* Harper Collins, 2009.

## CHAPTER 6

# India's Markets, Economy & Industry

The past sixty-eight years of independent India have been marked by major changes in the evolution of India's markets, industry, and financial structure.  Independent India was born in 1947 with many congenital handicaps from a long and oppressive colonial rule that severely restricted the formation of Indian businesses and industry, left a poor infrastructure with low connectivity, a nearly bankrupt economy, and a large population entrenched in unspeakable poverty and poor health.  Upon independence, democratic India adopted a highly government controlled, socialistic, and protectionist economy.  The results were mixed.  The country made substantial improvements in infrastructure in the form of a large network of roadways, a much larger system of railways, dams, hydel and coal power plants, and the like.  At the same time, many protective measures were adopted to provide a breathing space for the few existing Indian industry to rev up, for new companies to form, and for domestic economic engines to start up without being slaughtered by competition from very powerful, highly industrialized, exporting nations [1].

Noticeable improvements were certainly made in many areas, but the overall rate of economic growth and improvement was dismally low [2]. The high rate of population growth nullified even the limited progress that was achieved. Slowly, the country began heading for an economic disaster. While on the verge of bankruptcy, India took some bold steps in 1991 under the leadership of Prime Minister P.V. Narasimha Rao.  It ended the erstwhile permit *raj*, many tariffs and trade controls, and a variety of government regulations. Dr. Manmohan Singh,

India's then Finance Minister and the main architect of those policies, said, "We got government off the backs of the people, particularly off the back of India's entrepreneurs" [3]. The rest is a different history of India, the India that is now viewed as sporting an economy with the highest growth prospects.

There are many books that discuss India's economic history. Ramachandra Guha's history of independent India [4] is elaborate, but highly readable. The two books of Panagaria [5] and Subramanian [6] are particularly useful in understanding India's economy and its transition from stagnation to high growth.

The economy, market, and financial structures that provide the backdrop for Indian innovation together form a landscape that is highly fertile or depressingly arid, depending on one's perspective. My own attitude is that it is fertile, and even more so because its limitations do give ample opportunities for innovation. Let us start with a discussion of India's markets.

## 6.1 India's Markets

In the last two and a half decades, India's markets for goods and services have experienced a dramatic transformation and, so has the world community's perception of India. The opening of the economy by the government led by Prime Minister Narasimha Rao (1991-1996) was certainly the major turning point for the country and its economy. A subsequent government under Prime Minister Vajpayee (1998-2004) expanded the liberalization process even further. It privatized many state corporations including the Videsh Sanchar Nigam Limited (India's internet services provider), established special export processing zones and software parks, and initiated a variety of steps to woo Non-Resident Indians and their participation in India's development.

Above all, Vajpayee's bold steps of removing the nuclear ambiguity of India and asserting India as a nuclear weapons power that has mastered the complete cycle of nuclear technology resulted in a significant change in the attitudes of international think tanks and the major powers. As the following shows, those attitudes changed drastically from one of condescension, neglect, and bullying to a realistic one seeking cooperation for mutual benefit.

> "... (The) strategic decisions (of India) already in place will not be rescinded by any future government – even one composed of parties utterly opposed to the BJP. ... (The) 'International nonproliferation regime' is ultimately an American regime sustained by American power in the defense of what are primarily American interests. ... U.S. objectives must not consist of browbeating India into meeting certain political demands but must instead focus on understanding where India stands." – Tellis [7].

Helpful certainly in the process (Pakistan may note!) has been India's commitment to democracy and civilian control, responsible and restrained behavior, and active participation in world bodies like UNSC in positive ways.

The two successive multi-party (UPA) governments led by Prime Minister Manmohan Singh (2004-2009, 2009-2014) solidified this changed status of India further, particularly through building stronger ties with the United States. That government also improved its relationship with China, Japan, the European Union, and many of the developing nations. All these helped to drive the Indian economy into higher gear and resulted in the creation of a large 'middle class,' which now forms one of the largest markets for goods and services. Simultaneously, India too graduated to be the largest arms importer of the world (which, in my view, is a dubious distinction given that such imports divert valuable resources from national development and are highly detrimental to domestic innovation.)

### India's Consumer Market

The noted marketing scientist, Bijapurkar [8], asserts that the onus is on a market seeker to define what India's market is, for, there is no such thing as *the* Indian market.

> "India is not one entity but actually a kaleidoscope of culturally diverse, ethnically varied and linguistically distinct different Indias. Each of them is a mini country in its own right, existing in many eras." – Bijapurkar [8].

India's consumer market comprises a set of vastly different pockets exhibiting considerable variations in tastes, attitudes, and affordability. Some of the challenges (see [8]) India poses to a large foreign company wanting to sell its wares in India are also challenges for the Indian innovator in making a commercial success of his/her invention. Yet, they also yield a great opportunity for domestic innovation and invention. Unfortunately, these have not been explored systematically from that perspective by India's management *gurus* whose obsession has always been large corporations and, particularly, foreign MNCs. Indeed, it would be nice to see a change in the focus of India's management experts if only to stoke the engine of Indian innovation and to help India's innovators succeed with scale.

As India and the Indian economy have changed, so also have the attitudes and psyche of the Indian consumer. Marketing scientists segment India's consumers into a set of age-specific subgroups of which the following have distinctly different attitudes: the 'midnight children' born between about 1940 and 1970; the 'midway children' born between about 1970 and 1985, and the 'liberalization generation' born after 1985. The majority of 'midnight children,' like their predecessors, the 'pre-independence generation,' bore the cross of a highly impoverished nation. They were influenced by Mahatma Gandhi and various admonitions not to fritter away foreign exchange, and were typically frugal, *swadeshi* (driven by national self-sufficiency), and disdainful of conspicuous consumption. The 'midway children' had already started enjoying some benefits of economic progress and were more ready to throttle back on their reticence for consumption and for foreign goods, albeit with occasional feelings of guilt. Together, these two groups along with the pre-independence generation of India form about 45% of the nation. But the real bulk of India (nearly 55%) is part of the 'liberalization generation,' and the urban middleclass members among them are ready to spend with abandon and have no qualms about the source (foreign or domestic) of any of the goods or services as long as they can afford them (even if it be through a credit purchase at an usurious rate of interest). And, 55% of today's India translates to about 715 million people. About half of that are people above 15 years of age. Apart from these are the very affluent people of India who form a great market including for luxury items like expensive cars and SUVs; herein lies the lure of India as a market for the exporting countries of the world.

## Who Shall Benefit?

The trillion-dollar question for India is: who is going to capture and benefit from the large consumer and business markets that are developing in India? Will it be Indian innovators and businesses, or a conglomeration of foreign MNCs? If India's progress is to be sustained and is to move India into the realm of developed nations, obviously, it is essential that the giant opportunity offered by its internal market is not missed by India's own domestic commercial enterprises. If that opportunity were to be missed, it is certain that Indians will become cheap laborers enriching others, any progress that has been achieved will become evanescent through a rapid shift in trade balances against India, and, eventually, today's friends will flock to greener pastures leaving India to pay for its own sins of neglect and prodigality. That makes it absolutely essential for all stakeholders in India to take some necessary steps starting right now. A lesson coming out of China loudly and clearly today is that not developing and producing for one's own market is a prescription for eventual economic hardship. As I will argue below, innovation is a key ingredient to capture the unfolding opportunity.

Panagaria [5] provides an interesting and illuminating comparison of South Korea with India. He attributes the growth of that country mainly to its outward orientation. But his analysis, unfortunately, ignores some basic and important facts. For a while, South Korea had pursued a policy of high interest rates and other fiscal measures to attract enormous foreign investment into the country. South Korea's attempts to shine on foreign investment and through making goods for others only landed South Koreans in jobs yielding sub-human living standards [9], [10], and later to a major economic collapse due to sudden withdrawals of foreign investment [11]. The situation was further exacerbated as even cheaper labor became available in China. The resurgence of South Korea is really due to its transformation into a highly technologically advanced and innovative nation. Even before wireless communications, Internet, and Internet-banking saw significant penetration in the USA, they had already become quite popular in South Korea. The amount of research in South Korea on wireless technologies and the large number of patents being generated by South Koreans alone are astounding. Samsung has displaced the venerable Sony in the television market and Nokia in the wireless phone market.

Similarly, South Korean automakers are now capturing markets all over the world to noticeable levels with their highly improved vehicles. South Korea's presence in the steel market is also significant. This is particularly remarkable considering that South Korea is a very small country (with an estimated population of only about 49 million in 2015) and does not have a large internal market. It had to develop a global market for its products, and it did that mainly through innovating quality products and competing effectively based on quality and price. It should also be noted that South Korea makes a substantial chunk of its products in other countries. In short, what helped South Korea is not just the opening up of its economy, but even more its commitment to innovation and ownership of globally successful big corporations that benefit from those innovations providing a high margin of profits. That is *the* lesson that India must learn from South Korea. Liberalization may be an important ingredient to get reciprocity in opening global markets for Indian products, but it is certainly not sufficient by itself and may even be detrimental if it is not accompanied by other strong thrusts to innovate and own the innovations.

Let us, therefore, turn to some important areas where improvement is immediately needed so that India can leverage its own internal markets to the fullest extent. Needless to say, India must, in addition, expand its footprint in the global market as well.

## Making India Competitive

Indian consumers demonstrate a notable preference for foreign made products and consider Indian made products to be inferior in quality. While some of this preference is due to snobbery, a large part is because there are some real quality differences. Simple appeals based on nationalism and the importance of foreign exchange reserves alone are not enough to lure the Indian consumer to buy Indian made goods. Some concrete actions have to be undertaken to make Indian products competitive both in terms of quality and price. Such quality improvements will also position Indian products better globally. Implementing the following several steps can prove to be extremely advantageous.

*Standardization & Quality Control:* The government, industry, and various co-ops and associations of smaller manufacturers should come together to evolve standardization procedures to assure that the products made in the country have a uniform high quality. An 80-20 rule based approach identifying the most important sectors of the economy and working through a prioritized list appears to be a good approach to follow. See Chapter 3 for more on standardization and how it is used by developed nations. Indian manufacturers should make greater efforts to obtain consumer feedback and improve products and services constantly. Registries must be maintained for consumer complaints, and summaries and analyses must be used to identify trends and common patterns, and to drive corrective action in a timely and effective manner. Independent agencies, similar to the Consumer Reports of USA, should be formed and encouraged to perform independent product testing and issue comparisons, and their results should be disseminated widely to generate competition for greater quality improvement.

The safety levels of products deserve much greater attention in India than what they get today. Manufacturers of all sizes need to be educated to make this an important part of their product development. For example, many of the products, like most domestically made children's toys, would not pass safety norms of developed countries. Similarly, many locally made and sold stainless steel products like spoons and tumblers have sharp edges causing people who use them to suffer from cuts and bleeding lips. Many other such examples exist all around. No wonder, safety issues are frequently invoked by other nations, and often legitimately, to block entry of various Indian products [12]. Each substandard product manufactured in India sheds a bad shadow on all Indian products, including the good ones. It is, therefore, essential that the government assign the highest level of importance in enforcing quality and safety across the board in manufactured products. Constructive help should be offered where deficiencies occur due to lack of knowledge or lack of reliable processes; non-trivial and visible punishments should be administered when they are deliberate and due to profiteering.

*Tort Issues:* A major deficiency is the almost total absence of the means for meaningful legal recourse for consumers who are dealt poor quality or insufficient safety by negligent or bad businesses. The legal system needs to step up to the plate to provide quick remedies to consumers so that it forces quality and safety improvements. The present inefficient and corrupt law enforcement systems and the legal system with its inordinate delays have to be improved and also augmented by more effective systems.

*Scaling for Price:* Given the size of the Indian market, those who can achieve economies of scale can drive prices low and acquire the largest market share. This puts many Indian manufacturers and businesses at a terrible disadvantage. A multi-pronged approach must be followed to deal with this issue. One is to provide protections through tariffs and other means for a limited period of time until the Indian industry can become competitive. The second is to remove all regulatory barriers that get in the way of achieving scale. The third is to allow outsourcing/offshoring of manufacturing to other countries while India develops its own platforms for mass production and flexible manufacturing. Industry may need to be provided with low cost loans to set up such platforms and to amortize its cost over a longer period so that shifting to indigenous production (needed for mass employment) is not at the cost of significantly higher prices. These are areas where government can work with industry and develop policies and rules that can build the foundations for a strong economy.

*Probity and Transparency.* At the risk of repeating myself *ad nauseam,* I must once again emphasize the need for the highest levels of probity, transparency, and national consciousness among all involved in the evolution of new policies and their enforcement. Great plans fail in India at the implementation stage, and up-front safeguards are needed to prevent processes from being vitiated by corruption. Scams like those of Satyam Computer Services [13] and Ranbaxy [14] do not help India's image. Perpetrators should be brought to justice quickly, effectively, and visibly. As Indian industry becomes more competitive globally, it will be subjected to even greater scrutiny than it is at present and may possibly suffer even false attacks. It is, therefore, wise to be proactive.

*Privacy and Security.* Privacy and security concerns have become increasing challenges for India. The preponderance of foreign products, services, and players in high-tech areas like electronics, communications, and the Internet (and, particularly, social media) have exposed India to significant security risk and Indians to invasions of privacy. A strong need exists to develop domestic businesses in sensitive areas that can pose stiff competition to foreign players and encourage responsible behavior from them if only to maintain market share. The government should form public-private joint ventures in key areas and not allow the nation to be dependent entirely on foreign technology and services. That dependence also robs India of one of its important assets, namely its human capital.

## Distribution and Connectivity

In India, the biggest challenge for an individual inventor or innovator, or even for small and medium sized businesses, is the non-availability of efficient and affordable marketing and distribution channels. First and foremost, it would benefit innovators enormously if there were a body or set of bodies, both public and private, that could test and certify products with respect to safety and claimed functionalities and quality, so that new products that get introduced meet with less skepticism in the marketplace. There is a need for neutral bodies that can review a product, compare it against competing products, and also serve as a gating mechanism to block bad products. Similarly, systems to affordably market one's product to a large group of people, such as through a video-based shopping network, should be created. Many new products in the USA get introduced through television channels like QVC and HSN, which instantaneously create a large clientele for new products.

The advent of e-commerce is a boon to the small Indian businessman. It is necessary for Indian businesses to adapt quickly to that model. India's eagerness to allow 100% foreign direct investment (FDI) in e-commerce may accelerate development in this arena, but it does pose the risk that sites funded by FDI may provide preferential treatment to foreign products or may exhibit other types of bias detrimental to Indian businesses. There is a need to set ground rules up-front and to have effective monitoring of such sites. It also behooves Indian businesses to evolve domestic e-commerce sites at least through some cooperative consortium efforts and to quickly get a foothold for them in the public eye.

Transportation systems in India have to be improved significantly to increase market reachability for all types of industry. According to a World Bank study [15], "The service [for freight in India] is not adequate for the higher-value manufacturers or the time-sensitive export trade that comprises a growing share of the Indian economy." It is unfortunate that India, which manages, with considerable efficiency, a very large railway network, has failed miserably with respect to its postal service. The increased use of e-commerce should be seen by India as a new opportunity to revive and make efficient India's postal system and to use it as a catalyst in helping Indian businesses widen their markets. This could also be a major jobs creator for India!

## 6.2 India's Economy

The handbook of economic statistics [16] for the year 2014-15 published by the Reserve Bank of India has many tables providing a wealth of information on India's economy going back many years. I will touch only on some important aspects of the Indian economy as they relate to our main topic, namely, innovation in India and particularly capital formation to support a high level of innovation.

For an examination of the Indian economy up to 2006, Panagaria [5] is one of the most accessible sources for the statistics with interesting interpretations. The material in [16] can help one to quickly bring many of Panagaria's discussions up to date with the inclusion of more recent data. Panagaria has provided an objective and scholarly assessment instead of hyperboles chastised in the following comment of his.

> "A sure-fire way for an outsider to capture the attention of Indian audiences is to tell them that India is destined for stardom in the twenty-first century. If he can also argue that India's growth is more sustainable than that of China, and that it might even overtake the latter in the near future, the audiences may reward him with an instant celebrity status." – Panagaria [5].

My own interpretation of the macro-economic data of India leaves me ambivalent like Panagaria, and this is due to the mixed messages coming out of different economic measures. India, certainly, has the ability to move into the realm of developed nations, but it is doubtful if that is possible unless it pursues some specific economic policies that would form a major departure from those of the last many years. Let us start by examining India's foreign trade; the supporting data here is taken directly from [16].

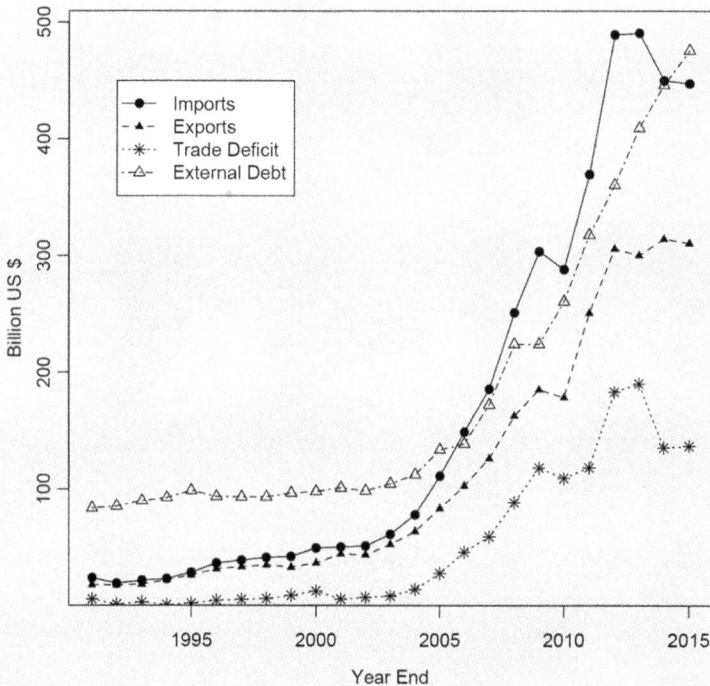

Figure 6.1: India - Foreign Trade
Imports, Exports, Trade Deficit & External Debt

From the curves in Figure 6.1 showing the total exports and imports of India that exhibit exponential growths, it is clear that India is getting more integrated into the world economy. However, it must also be observed that concurrently India's trade deficit has also widened significantly. More alarmingly, the total external debt of India has grown exponentially. This raises, in the language of [17], an important question as to whether, as a nation, we are getting more "income statement rich" while simultaneously becoming "balance sheet poor." The continuation of this trend is not in tune with our ambition to become a developed nation and to redeem fully our "tryst with destiny" [18].

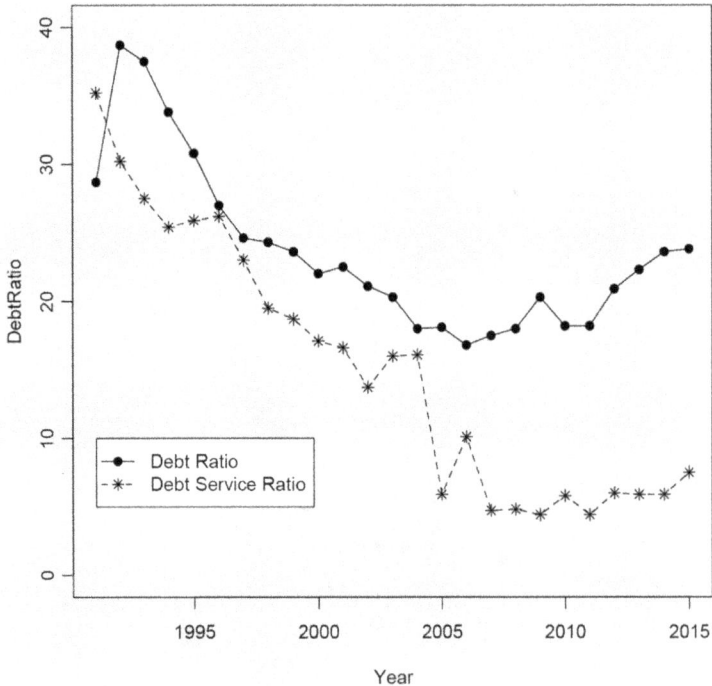

Figure 6.2: India's Debt Ratio and Debt Service Ratio

Turning to Figure 6.2, a first glance gives the impression that India's debt ratio (ratio of foreign debt to GDP) and debt service ratio (ratio of loan repayments and interest to GDP) have become favorable. [I have been unable to determine the reasons for the sudden drop in the debt service ratio in 2005 and the source of funding for effecting that.] However, when read in conjunction with the rise of foreign debt shown in Figure 6.1, there is certainly cause for concern. It should also be noted that these quantities are beginning to rise again during the last five years.   There is certainly a need for India to reign in its trade deficit as well as foreign debt.

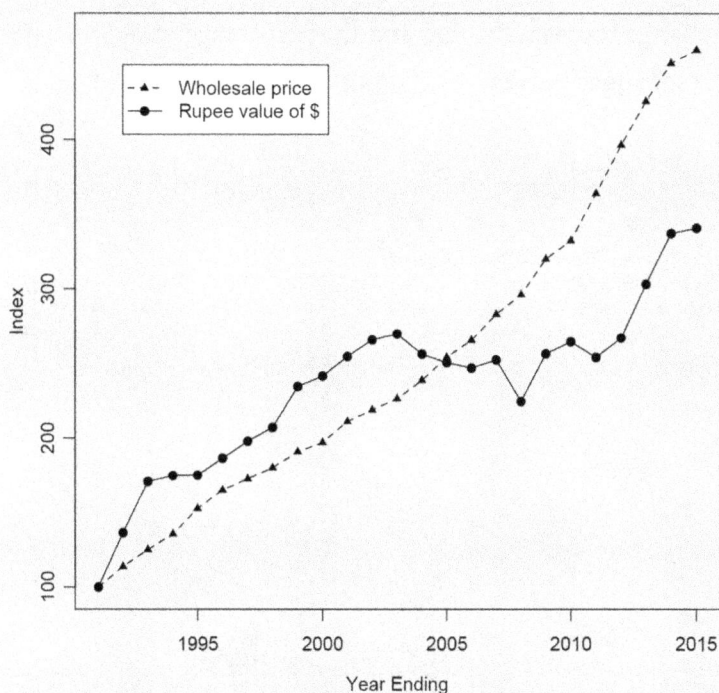

Figure 6.3: India Wholesale Price & Rupee Value of US $

In Figure 6.3 we have plotted the evolution of the wholesale price index in India (column AC in Table 40 of [16] converted to an index with base year 1990-1991 making the assumption that base changes have been made in a manner to make this possible) against a similar index for the rupee value of one US dollar.  The crossing of the curves and the divergence of the wholesale price index from the exchange rate index certainly deserve some attention. Does this, along with the widening trade deficit, portend yet another devaluation of the Indian rupee? Would there be more inflation? Would that engender a monetary policy that would adversely impact the cost of capital in India?

Our next focus of attention is on India's foreign exchange reserves. Presented below are two figures, Figures 6.4 and 6.5, presenting the foreign exchange reserves of India and the percentage of debt covered by foreign exchange reserves.

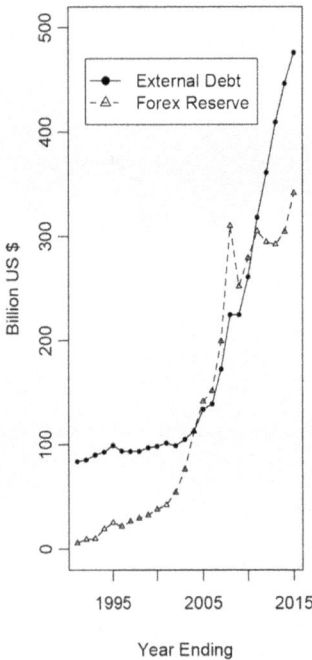

Figure 6.4: Exchange Debt & Foreign Exchange Reserve

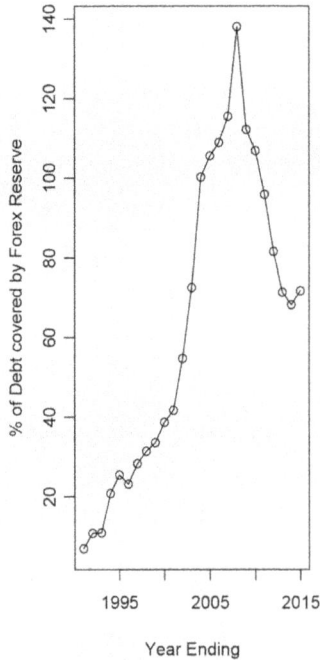

Figure 6.5: % External Debt Covered by Forex

Certainly, as can be seen from Figure 6.4, India has accumulated a significant foreign exchange reserve and has also managed to have a (short) period when the reserves exceeded the Gross Debt. That is quite encouraging. However, since 2010, the foreign exchange reserve as a percentage of Gross Debt has experienced a significant decline. Figure 6.3 clearly shows that this is due to increased indebtedness over the last five years.

An examination of Table 158 of [16] shows that only 24.2% ($115.16 billion) of the total debt ($475.81 billion) corresponds to NRI & FC deposits, and the remaining 75.8% is indeed borrowing by the country. Based on the macro economic picture above, it is imperative that India reduce its trade deficit, especially, by curtailing imports in avoidable areas and also reduce its foreign debt significantly. The risk of having to devalue the rupee again appears to loom large if the present situation is allowed to continue.

India's goal should be to shore up its balance sheet. Innovations along with ownership and marketing of new innovations are absolutely essential for India to build up its financial standing internationally. Innovations providing high margins are particularly needed, given the significantly devalued Indian rupee. A major impediment to investment for research and innovation in the country is available capital. Decreasing the dual burden of debt and debt servicing will provide more funds to support domestic innovations.

Over a third of India's import expenditure is for petroleum, crude and related products (36.6% in 2013-14). Along with increased investment in alternative energy sources like wind and solar power, India needs to also take steps to reduce its dependence on foreign suppliers of alternative energy products so that the net imports of energy related products do reduce drastically. The pharmaceutical industry in India has executed well in establishing a market for itself by identifying products that go off patents and become candidates for generics. India should replicate that kind of a success in key areas where its import expenditure is high so as to provide economical import substitution and also to create opportunities for export. Identifying such areas and products should assume a high level of importance in the nation and among its inventors.

India is extremely strapped for capital. Every avenue that can increase capital formation and create new domestic businesses needs to be explored. Several of these have been identified in our earlier discussion. These include reducing the lure of foreign made goods, making Indian goods competitive on quality and price, reducing the amount of "black money," unleashing unproductive passive assets such as those in the form of gold for gearing up the economy, and replacing the culture of high dividends and short term gains for long term investment and sustained growth. Most importantly, India needs to take steps to ensure that Indian companies capture the expanding domestic market so that they can expand in size, scale, and profitability and make the nation less dependent on foreign direct investment and foreign loans. That brings us to a discussion of Indian industry.

## 6.3 Indian Industry

Among the important factors affecting innovation discussed in Chapter 2 are many that deal with the state of industry in a country such as the size of its domestic market, its competitive structure, and its commitment to R&D. Clearly, a big domestic market has come to exist in India; however, it is not clear that India's industries are poised to benefit fully from it. While India's industries have grown enormously since independence, much is yet to be done to turn them into great engines of innovation to a level that would change India's global standing.

## Pampered & Protected

While the many decades of the 'license and permit *raj*' prevailing in India posed many barriers to it, Indian industry was, nevertheless, also pampered enormously by its government's various protectionist policies. At least in the first three or four decades of independence of the nation, they have functioned in a sellers' market and have not had to worry much about quality, customer satisfaction, and competition. The protectionist environment and lack of significant competition had also removed the necessity for much innovation by way of products and processes. For a long time, India has had nothing more than an infamous 'screw-driver technology' that could, at best, do a final assembly. The large incumbents have also had the great advantage of their legacy distribution channels and network, and that created a sufficient moat against internal competition. In many sectors, that has not changed much even now, although increasing pressures are now being felt due to foreign competition resulting from governmental policies of economic liberalization. Future game changers may very well be e-commerce and social media, the former by the instant market access it provides to even smaller players and the latter due to the speed with which customer angst can be disseminated. Yet, a cultural hangover from past policies does indeed exist within Indian industry, and its removal is necessary for India to become globally competitive.

Notable exceptions to this rule have been the industries affiliated with various strategic programs of the government. In fact, many of the leading innovations in India, including such things that have supported the most recent domestically developed GPS system [19], have come through such government involvement only. Another island of success has been India's pharmaceutical industry. Although some recent areas like software, information technology, and BPO in India may have gained much attention and may even be globally competitive, they have made hardly any noteworthy innovations and compete on cost only. For many such areas, since the arbitrage game they play will eventually end as labor costs in India increase, sustainability requires a highly accelerated level of innovation, new product generation, and a graduation out of mere body-shopping.

Whether it is the two centuries of colonial rule or the protectionism of the first several decades of independence, the overall culture is one that looks up to the government to provide solutions for almost everything. For India to progress on the innovations front, in the long run, it is essential that the dependence on government be lessened by industry. Indian industry must take steps to invest significantly in R&D and become highly innovative. In the short run, the government may, however, have to play a significant role in effecting changes. We present two ideas (the Fraunhofer Model and an R&D tax) later in the chapter in the context of accelerating industrial R&D and industry-university linkages.

## All in the Family

Most businesses in India, including some large public companies, are controlled by their founding families. While the amount of infusion of professional management is certainly increasing, yet family members hold many key positions, including on the boards, and decisions of strategic importance depend to a large extent on the whims of a small number of them. The lack of expertise in and knowledge of many modern technologies, business and management principles, and modern strategies of growth combined with risk aversion in these small groups of decision makers adversely affect these companies' level of commitment towards R&D and innovation. Both the companies and the investing public will have to take greater cognizance of the need for changes in this arena. Overall, Indian stock markets need to adopt valuation strategies more geared to growth than to income so that companies are incented to innovate and grow.

## Lack of Commitment to Scale

Both due to government policies and the lack of willingness to give up ownership by families, a culture of developing scale through acquisition has not taken deep roots in India. I have already noted how opportunities of achieving scale are missed by good Indian businesses as, for example, through standardization and franchising. Needless to say, scale is important to reduce unit costs and to increase margins and, thereby, generate more funds for investment in R&D and innovation for the future. Unfortunately, Indian industry is not geared to this and relies very much on organic growth.

Strangely, however, in the new areas of high-tech, one sees a contrary situation too - that of selling out too soon and often to a foreign player - which deprives the founders and the nation the full benefits of the innovation of a domestic venture. Until recently, many new ventures in India are funded by venture funds, which operate under the US model, and there may be a serious mismatch between the ventures' expectations and market conditions in India. That has actually led to the interesting question as to whether venture capital firms have adversely impacted Indian start-ups and innovation initiatives [20]. India badly needs a culture wherein the large Indian companies shall provide the necessary umbrella and support for innovative fledglings and not let them be lost to the nation through early dissolution or foreign acquisition. The government, through the provision of special incentives like longer term assurances with respect to policies, low cost

loans, and special tax treatment, needs to provide a climate where domestic large businesses can be made to adopt the role of supporting
and nurturing new ventures. There is also a need to consider lowering
the threshold for investors to participate in venture capital funds so
that there is a greater level of domestic participation with intent to stay
and build companies and not to exit at the first opportunity.

## Short-Sightedness

Indian businesses are very short term focused. None of them has
built an internal research organization of international repute. A small
few of the very successful ones, like the Tatas, have certainly shown a
commitment to research at the national level by instituting and supporting various academic institutions, but even they cannot be claimed
to have taken real advantage of research to further their business
through applied research and innovation. The exception, once again,
is the strategic government business area, which does pull in much of
academia and industry in design and development.

Many factors in India get in the way of a long-term focus supporting R&D. One is the practice of large dividend payouts that seems to
be preferred over investments of the returns back into growing the
business. In addition to the large discount rates imposed by the high
interest rates prevailing in the country, that practice may very well be
also due to the many policies of the government that curtail individual
companies from growing very big. Also, the lack of assurance of policies prevailing in a stable way after each election puts a choke on long-
term investments with deferred pay-offs. Thus, efforts have to be made
on multiple fronts to induce companies to invest more in R&D. An
important area where much work is needed is in creating a strong
linkage between Indian industry and Indian academia in a way that

the intellectual assets of the country can be put to greater use for commercial success by its businesses, and for the businesses, in their turn, to provide greater support for academic research.

## 6.4 Industry University Linkages

A major deterrent to commercial innovation in India has been the disconnect between academia and industry. Much of the blame for that can be laid on Indian industry's lack of vision and national consciousness to invest significantly in R&D. Most leading corporations in the US spend about 8% of their earnings on R&D for new products and systems. In certain high-tech industries, this percentage is significantly higher. Stock pickers pay attention to these numbers, and that, in turn, influences the pecking order of companies and stock prices. In India, however, most industries operate in a seller's market, and Indian industry, with few notable exceptions, has opted to ignore the long term benefits but, instead, concentrate on short-term profits. The Dalal Street culture of high dividends as a primary determinant of stock price, and the large discount rate that applies in India, have made things even worse. Be that as it may, even for their own long term survival and prosperity in this new world of globalization, Indian companies cannot continue in that mode. But unfortunately, no directional change is visible in India even in high-tech areas. Even major information technology companies like Infosys are bereft of a really credible research organization of international standing geared at inventing the future, although some of their luminaries loudly criticize even major Indian educational institutions and seek fame endowing foreign universities. The government should indeed consider stepping in with some new and constructive ideas to correct this bad situation. I propose two ideas for consideration.

## The Fraunhofer Model

The following is an eloquent description, in its own words, of Fraunhofer, the applied, non-profit, majority government owned counterpart of the theoretical Max-Planck Institutes of Germany.

> "Fraunhofer is Europe's largest application-oriented research organization. Our research efforts are geared entirely to people's needs: health, security, communication, energy, and the environment. As a result, the work undertaken by our researchers and developers has a significant impact on people's lives. We are creative. We shape technology. We design products. We improve methods and techniques. We open up new vistas. In short, we forge the future." – Fraunhofer [21].

I entreat the reader to watch the short video "Does Work!" [22], which describes some of the great products and technologies like mp3 audio compression that have come out of this organization. I also recommend Harvard Business School's case study [23] on Fraunhofer, which has a fascinating account of the history of German science as a preamble. India needs an organization of the type of Fraunhofer. India can create one if it chooses to, and it must – not just in structure, but also in accomplishments and contributions to the nation. I will only provide a brief summary of the salient aspects of Fraunhofer referring the reader to the Harvard case study [23] and other cited references for more details.

Fraunhofer fills two major gaps in R&D. The first is the funding gap in supporting targeted applied research that is necessary between basic research and actual product development. In most environments, the latter two are supported respectively by academia and industry, while it is the middle ground that becomes an orphan. Fraunhofer has many in-built mechanisms to accomplish that. One of them is the close integration with academia whereby many of its scientists are allowed to have dual affiliations with Fraunhofer and a university. Fraunhofer provides five-year contracts to selected university faculty who wish to commercialize their research along with a large research grant. Collaborations with universities also provide assistantships to students in universities and allow them to develop their dissertations at the institute with access to high-tech and high cost equipment and top-notch researchers with a practical orientation. Above all, a researcher who sets up a venture has the ability to return within two years if the venture is found to be infeasible. The occurrence of failure of new ventures coming out of Fraunhofer, however, has been kept down at a surprisingly low level of 10%, compared, for instance, with the 90% failure rate of start-ups in the USA.

One of the major problems solved by Fraunhofer is the reluctance of private industry to invest in research due to the risks a large capital investment entails – the risks associated with others' benefiting from one's research or usurping it through modifications and improvements, or mere technological obsolescence. Fraunhofer charges its customers only for variable costs and takes out a big burden off the hands of enterprises wanting to invest in R&D. Fraunhofer maintains close ties with industry, and both large and small companies regularly contact Fraunhofer for help with their technical challenges.

Indeed, a striking feature of Germany is that the so-called *Mittelstand* of small and medium enterprises contributes significantly to Germany's advancement in technology through R&D investments. It accounts for 21% of Germany's exports and 9% of the total R&D expenditures in Germany. A 61% of these enterprises are involved in some innovation activity, while 51% of them have introduced organizational and/or marketing innovations. Many engage in contract research. In fact, a significant part of Fraunhofer's revenues comes from research contracts with the *Mittelstand*. India should try to mirror this with its own small and medium sized companies, encourage them to invest more in themselves through R&D, and provide the necessary support structure through R&D organizations that charge them only for variable costs and not for the embedded infrastructure costs.

The organization of Fraunhofer is also highly conducive to creative freedom. It receives 65% of its funding from the government and the rest from industrial collaborations. Its sixty institutes are organized into five groups falling under major areas of importance. The central administration only sets strategic directions and identifies broad areas of research with the greatest innovation potential. Each institute is run autonomously, interacting among members of its own group and setting up partnerships with universities. The overall environment is one that keeps the best scholars wanting to come to it and stay there.

India's CSIR institutes resemble Fraunhofer in some ways in that they have a large budget and a large staff. However, they are no comparison to Fraunhofer, especially with respect to levels of new product creation or commercialization of research. This may be due to several reasons: hierarchy and central control; greater use of its resources by the government; strategic and defense related research; lack of a large venture capital under its control; absence of metrics that reflect commercial value; and a culture of rewards not fully based on merit and real accomplishments.

There is a strong need to either evolve CSIR more along the lines of Fraunhofer or to create a new organization whose primary focus will be industry and the commercialization of research. Such an organization should measure its success not in terms of published papers, patents applied for or obtained, and various human resources related social goals it satisfies, but, more importantly, in terms of the real wealth created for the nation through new products and the commercialization of research. The organization should function in such a way that it would provide real opportunities for enterprising university faculty wanting to enter the realm of commercializing their research and for industries wanting to gain from India's assets in academia and basic research.

## An R&D Tax

The second idea I propose for increasing the commitment of Indian industry to R&D is the imposition of an R&D tax on Indian corporations with revenues above a certain threshold. That tax should be a percentage of total revenues (say 2% to 4%). I say revenues because, unlike net earnings, this is a number that cannot be manipulated easily by creative accounting. A significant percentage of the taxes levied on a corporation in a year can be credited back to that corporation towards its tax liability for the subsequent year if the corporation shows demonstrable internal investment in R&D for the current year, and demonstrable new patent filings, products, and processes over each three-year rolling window. Those failing to demonstrate such involvement in R&D should forfeit the R&D taxes paid. A significant part of the forfeited amounts should go into a common pool supporting academia-based applied research and prototyping. Its results may later be licensed to industry based on a competitive bidding process so as to stoke the engine of research and innovation even more. A portion of the common fund should also be accessible to private research institutions, including in industrial houses, on a competitive basis. The entire program should have significant involvement of scientists and technologists and be run with the highest levels of transparency so that it does not fall victim to corruption that is, today, India's largest curse imped-ing any real change.

The above proposal will certainly raise eyebrows in many quarters. Therefore, let me cite two simple examples taken from a Harvard Busi-ness Review article ([24], pp. 27-42) to illustrate the need for some dras-tic intervention. We have, in India, some very large industrial houses that have large medical divisions. As a country, we also have some of the brightest engineers and engineering faculty that industries can turn to for help. No educational institution is prevented from working on problems important to its field, nor does it suffer from a lack of mon-ey for academic research. Yet, why is it that it has to be a GE from the USA, which has to invent a thermal blanket for preventing infant deaths in India due to cold conditions even though, every year, we read in the newspapers about the deaths of many due to cold waves in North-ern India? Similarly, we all know that access to hospitals with large-scale diagnostic equipment is a serious problem for the rural poor in India. Yet, why is it that it is a foreign industrial house that has to make portable, small, ultrasound monitors specifically for markets like India? That these are done with Indian engineers in India and are now becoming global products in a strategy called 'reverse innovation' should be viewed by India as adding insult to injury, although no one but we Indians should bear the blame for the missed opportunities to enrich ourselves and our research in meaningful ways. One would be amazed to see the amount of research that goes on in US universities, specifical-ly targeting the needs of China and India that are perceived as the biggest markets of the future. Why is it that we have so little going on in our own universities and elite institutes? More importantly, how can we change this scenario?

Without some active arm-twisting by the government, one does not see noticeable changes happening in Indian industry's attitude to serious and purposive R&D related activities of a credible type. The majority of the Indian corporations have shown a tremendous incapability for innovative and scientific vision or national consciousness in this important arena. It is time for the government to step in so that 'Make in India' does not become a new form of a screwdriver technology that lends only cheap labor, but translates into 'Invent in India and for India' in every sense of the term.

## Visible Encouragement

Greater recognition is needed for science and engineering as applied disciplines for generating research and innovations that can improve the national economy and quality of life. The higher glory given to theoretical work and theoretical researchers in the choice for various national awards and recognition has the effect of driving the brightest away from applied work that is badly needed. Even as it celebrates Indian achievement in the areas of publications and foreign awards for theoretical advances, the Indian media needs to pay greater attention to applied research and curtail its fascination for the highly theoretical and esoteric forms of research. As a people, we Indians need to give much higher priority to celebrating research of a type that occurs in India enhancing India's economic progress in measurable ways. Harping on past glories has to cease as a pastime replacing the lacuna in our current performance. Also, achievements of NRIs abroad may have a significant role in building our confidence as capable scientists and engineers, but, by itself, it does not help India advance. A greater focus must be on drawing the NRIs as a resource for building our own capabilities as a nation wanting to advance in research and its commercialization.

## The Start-ups Fever

It is heartening to note that India is launching major efforts to create a huge host of start-ups [25]. There is no doubt in my mind that it will result in highly visible successes and many notable innovations by Indians in India. But if the main driver of those is Foreign Direct Investment (FDI), then how much of the benefits of these will accrue to India, and what fraction would it be of the total value generated? Will there be enough Indian equity participation and ownership to assure that India's share will indeed be sizable?

Take the recent news that the Indian government has allowed 100% FDI in e-commerce [26]. I welcome that as a wonderful piece of news given that it could help many Indian small and medium businesses acquire access to a much larger market instantaneously. But will it really do that? [Already, Flipkart, it appears, would not allow a merchant to use its services without a minimum portfolio of ten products, a policy that would clearly exclude many small innovators and businesses.] Will there be mechanisms and oversight to make sure that the e-commerce enterprises formed with 100% FDI shares will not gravitate, intentionally or unintentionally, to provide preferential services to foreign suppliers of products? If that turns out to be the case, wouldn't Indian businesses and the Indian economy be in even worse shape? I do not believe my fears to be unfounded given that similar concerns are ruling high in the EU [27] in comparable areas.

Shining on borrowed feathers is never a good policy. May be, the ball that India wants to dance in and its dress code require her to borrow some feathers at present, but may it not be forgotten that the strategy is not good for the long term! We need a more sustainable strategy based on innovations and owning what we create. The stakes and responsibilities associated with evolving a more sustainable strategy should not concern only India's politicians but also its policy makers, academics, scientists, and businesses. Each one of us needs to step up to the plate to ensure that India shall have her own stunning wardrobe, nay an enviable boutique, or we may have to settle for a situation even worse than what it is today.

## SUMMARY & POSSIBLE ACTION ITEMS

**Secure the Domestic Market**: Necessary steps must be taken to make sure that the growing domestic consumer market is captured by Indian industry. This involves: (i) improving quality of Indian products through standardization and certification processes; (ii) improving the availability of objective product and customer satisfaction reviews (iii) facilitating scaling by industries to achieve cost competitiveness; (iv) reducing the lure of foreign made products based on cost and quality; (v) improving the availability and ease of distribution channels, particularly e-commerce and television based sales.

**Import Substitutions**: Steps should also be taken to reduce India's imports burden. India must distinguish between imports that are of an investment type versus imports that must be treated as expenses only. It must move R&D towards creating domestic substitutes by identifying products losing their patent protections and also creating new products that are patentable in their own right.

**Indian Ownership**: It is not enough just to innovate, but even more important to own the innovations and to benefit fully from them. Steps must be taken to ensure that the intellectual and human capital of India benefit India more than other countries and their corporations.

**Improve India's Balance Sheet**: Effort must be made to reduce the burden of foreign debt and debt servicing that necessitates repeated de-valuations of the rupee and the maintenance of a high interest rate impeding R&D investment, business creation, and business activity. Government deficits must be reduced. Wealth and income reporting as well as tax collections must be improved through better and effective enforcement.

**Accelerate R&D by Industry**: It is important to consider the imposition of an R&D tax on large corporations that can be credited back for demonstrable R&D that leads to new products and additional revenue. Creating or evolving an existing one into an institution of the type of Germany's Fraunhofer to promote greater applied research and commercialization of research can be very beneficial.

## Notes and References

[1]   Carol Hansa: "Foreign Relations Without the Begging Bowl," *The Christian Science Monitor*, September 9, 1982.   http://www.csmonitor.com/1982/0909/090930.html

[2]   Meera Siva: "What's a Hindu Rate of Growth," *Business Line, The Hindu*, June 8, 2013.
      http://www.thehindubusinessline.com/portfolio/technically/whats-a-hindu-rate-of-growth/article4795173.ece

[3]   "India Escapes Collapse," Commanding Heights : The Battle for the World Economy, Public Broadcasting Service (PBS) Video Excerpt, https://www.youtube.com/watch?v=uuPKtTEPQB8

[4]   Ramachandra Guha: *India after Gandhi: The History of the World's Largest Democracy*, Harper Collins Publishers, 2007.

[5]   Arvind Panagaria: *India: The Emerging Giant*, Oxford University Press, 2008.

[6]   Arvind Subramanian: *India's Turn: Understanding the Economic Transformation*, Oxford University Press, 2008.

[7]   Ashley J. Tellis: *India's Emerging Nuclear Posture: Between Recessed Deterrent and Ready Arsenal*, RAND Corporation, ISBN:0-8330-2781-6 (pbk), 2001.

[8]   Rama Bijapurkar: *We Are Like That Only – Understanding the Logic of Consumer India*, Penguin Books, India, 2007.

[9]   Matt Wilsey & Scott Lichtig: "The Nike Controversy," https://web.stanford.edu/class/e297c/trade_environment/wheeling/hnike.html

[10]  "Nike Sweatshops: Behind the Swoosh," https://www.youtube.com/watch?v=M5uYCWVfuPQ

[11]  William C. Hunter, George G. Kaufman, G & Thomas H. Krueger: *The Asian Financial Crisis: Origins, Implications and Solutions*, Kluwer Academic Publishers, 1999.

[12]  "2100 Batches of Indian Goods Including Those Made by HUL, Britannia, Nestle, Haldiram Fail to Gain Entry in US," *The Economic Times*, June 13, 2015.

http://economictimes.indiatimes.com/news/economy/foreign-trade/
2100-batches-of-indian-goods-including-those-made-by-hul-britannia-
nestle-haldiram-fail-to-gain-entry-in-us/articleshow/47649906.cms

[13] "Satyam Scam: All You Need to Know About India's Biggest Accounting Fraud," *Hindustan Times*, April 9, 2015, http://
www.hindustantimes.com/business/satyam-scam-all-you-need-to-know-
about-india-s-biggest-accounting-fraud/story-
YTfHTZy9K6NvsW8PxIEEYL.html

[14] "Ranbaxy whistleblower reveals how he exposed massive pharmaceutical fraud," *CBS This Morning*, November 6, 2013,
http://www.cbsnews.com/news/ranbaxy-whistleblower-reveals-how-he-
exposed-massive-pharmaceutical-fraud/

[15] India: Road Transport Service Efficiency Study, Report 34220-IN,
Energy and Infrastructure Operatons Division, World Bank, Washington, DC, 2005.

[16] *Handbook of Statistics on Indian Economy 2014-15*, Reserve Bank
of India, Sep 16, 2015. https://www.rbi.org.in/scripts/
AnnualPublications.aspx?head=Handbook of Statistics on Indian
Economy

[17] Thomas J. Stanley & William D. Danko: *The Millionnaire Next Door:
The Secrets of America's Rich*, Pocket Books, NY, 1996.

[18] Jawaharlal Nehru: "A Tryst with Destiny," Speech given on the dawn
of India's independence, 1947.
http://www.theguardian.com/theguardian/2007/may/01/greatspeeches

[19] "India gets desi 'GPS': 10 things to know," *The Times of India*, April
29, 2016. http://timesofindia.indiatimes.com/india/India-gets-desi-
GPS/listshow/52029512.cms

[20] V. Sivaraman: "Is venture capital killing India's startups?" *Quartz
India*, Nov 16, 2014. http://qz.com/297245/is-venture-capital-killing-
indias-startups/

[21] Fraunhofer, http://www.fraunhofer.de/en.html

[22] Profile of the Fraunhofer-Gesellschaft,
https://www.youtube.com/watch?v=Y_6oNUsA06k

[23] Diego A. Comin, J. Gunnar Trumbull & Kerry Yang: "Fraunhofer: Innovation in Germany," HBR Case Collection, March 2011 (Revised March 2012). http://www.hbs.edu/faculty/Pages/item.aspx?num=40275

[24] Jeffrey Immelt, Vijay Govindarajan & Chris Trimble: "How GE is disrupting itself" in *HBR's 10 Must Reads On Innovation*, Harvard Business Review Press, Boston, MA, 2013.

[25] "Modi announces tax Sops, funding support for startups," *The Hindu Busines Line*, Jan 16, 2016.
http://www.thehindubusinessline.com/economy/pm-modi-announces-tax-relief-other-sops-for-startups/article8114334.ece

[26] "Government permits 100 per cent FDI in e-commerce," *The Hindu*, March 30, 2016.
http://www.thehindu.com/business/Industry/govt-permits-100-per-cent-fdi-in-online-market-places/article8409495.ece

[27] "EU Deepens Antitrust Investigation into Google's Practices," *The Wall Street Journal*, Aug 21, 2015.
http://www.wsj.com/articles/eu-deepens-antitrust-investigation-into-googles-practices-1440178863

# CHAPTER 7

# India: The Government

India is the world's largest democracy and has remained so continuously since its independence in 1947 (see [1]). This is by no means a small achievement given the diversity of India, the "fissiparous tendencies" (a frequent phrase of Nehru) of various subsections of Indians at various points in time, and the machinations from outside (some of which still go on [2]). Since 1947, the governing parties and the prime ministers have changed at the central level in India many times. So have the governments in the states of India. But almost always, the change in leadership has been achieved in a peaceful manner through elections in which voter participation has been quite high. This is noteworthy considering that nearby countries have done very poorly on this score, some despite being blessed with notably less diversity. Contrary to the dissent and the shrill cries of the far left in India, India has certainly enabled its poor people to have a significant voice in government affairs and public policy. Unlike in affluent societies [3], the poor and the minorities have a major say in how they are governed, and the state has done much to uplift them.

The successive governments have done splendidly in terms of infrastructure development in the nation and the creation of a vibrant set of industries. A number of great educational institutions have been built in the country, and their graduates have made significant contributions to India, particularly in the strategic areas. In several areas, India has joined the small groups of countries that have achieved significant milestones in science and technology, the latest being the deployment of its own geographic positioning systems [4]. A major driver of all that progress has been, without doubt, the Government of India.

Despite the above commendable facts, the functioning of the governments at the central and state levels in India is in need of drastic improvement in certain key areas for India to become a developed nation. It is even more so given that the ethos of the country is one that looks up to government for almost everything. As many specific ways in which government can help to accelerate innovation and economic progress have already been discussed in earlier chapters, I will focus here on the big picture, and that too only briefly. But, before that, I must list a fundamental change in attitude that must occur in governance in India without which nothing else would come easily. So, let us turn to it first.

## Government, the Ruler

If there is one word I would like to see expunged from the Indian lexicon, it is the word "rule." In India, this is the operative word for the press, the media, and even the common man and woman to describe the government, the governing party, the prime minister, and any chief minister or minister at the central or state level. Thus, just as India has been 'ruled' by the Moguls, the various despot kings, and the mighty British, it has also been 'ruled' since independence by so many parties and politicians of all ilk and convictions.

The word 'rule' would be quite innocuous if only the government (both elected politicians and civil servants) in India were to act as people chosen by fellow men and women to *serve* them, and if they did not act as rulers ready to display their power and pelf at every opportunity. That tendency of the 'rulers' at the top, including the amassing of obscene levels of personal wealth at public expense, has indeed percolated all the way down to the petty peons so much so that India's government is, in many respects, more of a ruler than a service provider, a controller than an enabler, and an obstructer than a helper. Among others, it seriously affects the country's economic progress and the utilization of its crème of the crop. As noted in Chapter 2, it is imperative for government at all levels to adopt a different posture if the country is to progress rapidly and become a developed nation. That is, particularly so, with respect to innovation and business success.

For finding resources for social purposes, to provide greater levels of equity and justice for all, for protecting individual rights and possessions, to maintain law and order, and for many other noble purposes, the government certainly has to pass and enforce various laws and regulations that restrict people's freedom. In a democracy, the people trust some with these difficult and necessary tasks, and people have a right to expect that those so trusted shall honor that trust. The role of every government official should be to serve people to the fullest possible extent. Those vested with enforcement powers should start from the premise that one is innocent until proven guilty. It is also the duty of the government to bring the guilty to justice expeditiously and through fair and transparent processes without showing any favoritism or bias. The government should function as the lubricant that removes friction of all kinds that can impede economic progress or abridge the quality of life. In all these respects, unfortunately, India's governments, both at the central and state levels, would not merit even a passing grade. Government, after all, is made up of people, and a large number of people in government in India need to change and change enormously if India is to move forward.

Many of the ills afflicting the nation – like pervasive corruption, obstructive bureaucracy, red tape, frequent breaches of law and order, political violence, destruction of public property, and low productivity – map directly to the acts of omission and commission of India's 'rulers.' The progressive and alarming decline in efficiency and probity that has occurred in the nation has to be reversed, and every avenue explored to improve the functioning of governments in all regions and at all levels in India. That having been said up front, let us turn to some specific steps that Indian government can take to accelerate the country's progress through the act of empowering Indian innovation and India's businesses.

## Control Corruption

At various points in earlier chapters, I have lamented about the high level of corruption in the nation that vitiates most efforts at improvement. The leaders at the top can make grandiose plans with the highest vision, but none of that will come to fruition unless corruption at the lower ranks is fully eliminated and those responsible for implementation of policies and enforcement of rules are made accountable through appropriate rewards and punishments. Unfortunately, corruption has become endemic in the nation and is driving it towards self-destruction. Eliminating corruption has to be given the highest priority by the center and also the state governments in India.

Unfortunately, the enforcement agencies and the courts, which should play the most important role in removing corruption, seem not at all immune from that very malaise. Nor do they enjoy the ability to deal with corruption without political interference. There is urgent need to remove all forms of political interference in matters of law enforcement and to stop the protection of the affluent or people affiliated with the 'ruling party' from legitimate enforcement actions. Greater use of technology including audio and video surveillance, and easier admittance by courts of evidence gathered by their use, alone can reduce official and political corruption in the country significantly.

Tough legislation against corruption must be brought at the national level, with the central government being given the ability to indict and try the guilty even if they are primarily affiliated with a state government. As a deterrent, an urgent need certainly exists to make examples of some corrupt "leaders" with high visibility, as well as members of the general public who have been caught on camera indulging in violence and destruction of public property. The government should place very high priority on altering the (justified) perception that mobs and the rich often get a free reign in India, and that with money, political clout, and the threat of violence, anything and anybody can be bought in India.

## Improve Law Enforcement

Corruption and poor law enforcement go hand in hand and sustain each other. In India, the inadequacy of the court system as well as of well- trained and well-equipped police forces is well documented. Many open appeals have been made repeatedly for improvement through the provision of resources both from within and outside the court system [5] but have gone unheeded. The most recent passionate appeal [6] from the current Chief Justice of India is indicative of the level of desperation from overload felt by the Indian judiciary. The government needs to take immediate steps to increase support for the courts (and law enforcement agencies in general) and also set up alternative forums besides courts for redressing legal disputes. For businesses and innovation to thrive in India, it is essential that property rights are respected and become enforceable expeditiously and dependably within its judicial systems.

One should, however, note that the judiciary itself is not free from fault. Judges have failed to exercise their power to bring timely closure of cases and have given extensions and appeals freely allowing lawyers and defendants to game the system. They seem to have also developed a habit of dealing with the rich and the powerful with kid gloves even when the latter are accused of serious violations. Rumor of corruption among judges, at least at the state level, is rampant. A small number of Supreme Court judges too have been accused of serious corruption. In spite of these, it is fortunate that the courts still enjoy considerable respect from the general public, and before that respect erodes significantly, it is important to strengthen the courts so that overall respect for the law improves in the nation.

In the case of India's police, one sees both excessive shows of force and total incapacity to deal with mob violence. The significant causes behind these are: (a) inadequate number of police personnel; (b) poor training especially in handling mobs; (c) political interference; (d) the lacuna in the legal and court systems. The central government needs to enact some basic norms of performance by states that have the responsibility for maintaining law and order. The center should have the authority to intervene if these norms are not met. The present situation of the imposition of President's Rule in a state is an extreme step that comes into play only in extreme circumstances, and the big gap in oversight of the states in the discharge of their responsibilities has resulted in a terrible decline in law and order across the board with some states verging on intolerable lawlessness.

## Restore Merit

It is the duty not only of the government, but also of every Indian to commit to uplift all and to ensure that gender, caste, religion, and economic status do not become barriers to any individual in reaching his or her full potential. Society also has the obligation to repair past ills and to provide necessary help to train and equip disadvantaged people, even if they are defined by criteria other than economic status alone, to compete effectively and successfully. But the nation cannot afford to lower standards and to provide entitlements to any group. It is also necessary to ensure that help reaches the truly deserving.

The world is becoming increasingly competitive due to globalization, and countries that do not foster and uphold a culture of merit are bound to slip to the bottom. Any action that gets in the way of maximizing the benefits that can be derived from the meritorious in the nation will have a serious impact not just on them but even more on the people who need to be helped. So, if India is to become a developed nation and achieve a good standard of living for most of its citizens, it is imperative that steps are taken by the government and all political parties to come to a consensus to gradually eliminate current systems of entitlements and reservations and to replace them by a system of targeted assistance to the well deserved to compete effectively in a merit based system. This certainly is a collective responsibility of the nation and of each individual in it, and not just of the government. The opposition parties also owe it to the nation to foster a culture of merit and not to use entitlements and doles as vote getting tools. The political parties should evolve a common approach that will meet the dual objective of upholding merit and simultaneously helping the disadvantaged.

## Involve Specialists

There is a pressing need for a technical service cadre in government in several different areas where technology or specialized services have to be managed. In areas like education, R&D, and innovations management, the government should lead efforts through personnel with relevant technical and managerial expertise. The current practice of using generalists in all kinds of roles is not conducive to evolving best policies and practice. In the short run, a team-based approach that involves specialists with necessary skills and experience should be implemented for obtaining reliable outcomes. In parallel, it is necessary to make the personnel more accountable in terms of measurable metrics including meeting targets in a timely fashion. Evolving mechanisms in government for promotions and rewards based on demonstrable performance instead of seniority is also needed.

## Strengthen Education

India's challenges are so big that it is unrealistic to expect dramatic transformations to occur overnight. The real future of the nation lies in shaping the young generation well, and education plays the most important role. The government should give high priority to education reform. Once again, there is the need to bring national consensus in a way that standards can be established and administered at the national level while at the same time allowing individual states to meet their needs to nurture their local language and culture. The state governments need to balance state level requirements against the needs of national integration and free movement of the people and ensure that parochialism does not become an impediment. Above all, India needs to take the steps necessary to make sure that India's investment in higher education yields significant benefits to India. India needs to ensure that its crème of the crop does not leave India, but stays and contributes to India's growth. Many specific steps have been put forth for consideration in Chapters 4 and 5 to make education at all levels to be more effective. It should be clear from those discussions that an incremental approach will not provide the type of improvements needed, and that a thorough overhaul of the entire system is necessary. The government should lead the effort for comprehensive education reform and ensure that good quality education does not become the prerogative of only the more affluent.

## Support Domestic Businesses

India's economic policies and practice should be tuned to enable a high rate of growth by all types of domestic businesses. Specific areas we have noted earlier involve: (a) improving the distribution system, especially for small and medium sized businesses; (b) improving the transportation and goods movement systems within the nation; (c) ensuring government coordination and help in standardization and quality improvement; (d) guaranteeing a concerted involvement of the government in research and development leading to new products and processes; (e) increasing the availability of capital for businesses; (f) refining taxes so as to induce businesses to develop a long haul perspective and greater investments in R&D; (g) creating national institutions to promote interactions among industry, academia, and the government in specific areas of applied research and the commercialization of research.

## Increase Productivity Levels

In the context of improving the economy, the government needs to take some specific steps in increasing the productivity levels in all sectors of the economy. Specifically, government departments and em-

ployees need to be measured using metrics that correctly reflect their
true productivity and not just the time spent on the job. India has too
many paid holidays and vacation days compared to that of most developed nations, and these must be cut back.

Strikes and labor disputes reduce productivity significantly as do
demonstrations by various political groups. The government and the
courts need to step in to resolve them quickly. While the government
has a responsibility to ensure that labor is afforded fair treatment,
various regulations protecting labor – particularly at state level – have
had the negative consequence of significantly reducing accountability
and productivity. Simultaneously, we also see extreme exploitation of
labor in many sectors such as mining. There is a need for comprehensive reform in India's labor laws and their enforcement so that a proper balance is struck between protecting labor and ensuring accountability and productivity.

## Maintain Continuity

Business and the stock markets all over the world, including in developed nations, hate nothing more than economic and political uncertainty. It is absolutely necessary for Indian businesses to have a
certain sense of stability and continuity in government's economic and
fiscal policies for them to take a long haul perspective and to invest in
research and development.

One of the inherent defects of democracy as a form of government
is the natural tendency of political parties and elected representatives
to use their position in government to support what is popular than
what is necessary. In India, this tendency manifests itself often in the
form of giving out doles and entitlements to large sections of people at
the expense of businesses and individuals that are productive and benefit

the economy. In addition to reducing the level of individual accountability and initiative, the doles negatively impact the economy in the form of increased government deficits and debt. Indeed, many states in India are nearly bankrupt due to such policies of their governments. Therefore, greater effort has to be taken to reign in public spending. Controlling various macro-economic parameters like interest and discount rates should assume much higher priority than they are afforded today. Wherever possible, governments should replace welfare programs with workfare programs so that economic assistance goes hand in hand with an increase in the GDP and national productivity.

## Increase Self-Reliance

For bringing quick improvements in certain metrics like employment levels and greater availability of consumer goods, India should not bring in too much of foreign investment and foreign players at the cost of domestic businesses and domestic ownership. Just as for individuals, so also for the nation, the ticket to wealth is not through being employed by others, but through investments and ownership. The bigger challenge for the nation is to build a sustainable growth rate and an increasingly stronger balance sheet. In that context, it is imperative that national policies are geared to increasing domestic ownership of both businesses and new innovations.

In his book on India's nuclear policy, Karnad [7] recalls Homi Bhabha characterizing foreign inputs to India's nuclear programs as "booster-assisted take-offs" and his apt admonition, "A booster in the form of foreign collaboration can give a plane an assisted take-off, but it will be incapable of independent flight unless it is powered by engines of its own." The governments, both at the center and in the various states, need to develop plans to make sure that even as we collaborate with other nations and nationals to accelerate economic activity in our nation, we are building our own engines of growth.

India, and particularly its various state governments, should tread with extreme caution on mega projects peddled to them by foreign multinational entities, banks, and consultants. The warnings coming from the tales of Perkins [8] are such that every such project should be vetted thoroughly by independent experts. Given the real economic and environmental challenges faced by India, the priority of the nation and the states should be on products and projects of high economic and quality-of-life value and not on those that are for show. As an example, consider deploying bullet trains between some cities (as the one being mooted between Mumbai and Ahmedabad). This requires to be properly compared through a cost-benefit-analysis with alternatives to improve congestion and the levels of pollution in the cities through the provision of more intra-city facilities for reliable mass transport. It is fair to ask whether the provision of better and cheaper facilities for video conferencing over high-speed networks that can reduce the need for travel would not be more efficient than bullet trains. Similarly, it is necessary to examine how using the resources to improve the distribution and transportation network for goods by small and medium sized businesses compare with the provision of bullet trains that only the top few in the nation could possibly afford as a luxury. Rather than responding to various proposals originating from outside, the approach should be one of developing our own list of prioritized needs and searching for appropriate solutions. Incurring debt for a wise investment with returns far above the cost of the loan may be wise, but it certainly is foolish when it cannot be justified and is for expenses without high returns.

India is fortunate to be positioned well in many respects. It has the benefit of size and abundant natural resources, a large infrastructure built with patience and planning, and, above all, a demographic profile that is conducive to economic growth. Right now, India is enjoying a window of opportunity in many respects, and it is essential that the focus be on seizing that opportunity fully. The focus of the government at all levels should be to make India's balance sheet rich and also to ensure an equitable participation by all sections of society in that richness and growth.

## Notes and References

[1] "Emergency: The Dark Age of Indian Democracy," *The Hindu*, June 27, 2015. http://www.thehindu.com/specials/in-depth/the-emergency-imposed-by-indira-gandhi-government/article7357305.ece

[Starting June 1975, Indira Gandhi imposed for eleven months an emergency rule in India and assumed authority to rule by decree. Although some Western scholars and media have attempted to characterize it as an act of suspending democracy, it must be noted that those steps were well within the provisos of the Constitution of India. Unlike two Presidents of the US who got off easily for felony offenses, she was convicted by an Indian court of law, albeit for certain technical violations of election rules, and was forced to resign. Subsequently, her party suffered a major loss in the general elections. These events, if any, only assert the strength of democracy in India even during that difficult period. Indeed, objective analysts - even though they may strongly disapprove the actions of Mrs. Gandhi and especially the many abuses by those around her - must acknowledge India to have had a continuous and unflinching commitment to democracy since its independence.]

[2] Rajeev Malhotra & Aravindan Neelankandan: *Breaking India: Western Interventions in Dravidian and Dalit Faultlines*, Infinity Foundation, Princeton, NJ, ISBN: 978-1-937037-00-0, 2011.

[3] John Kenneth Galbraith: *The Affluent Society*, ISBN 0-395-92500-2, Mariner Books, The Mass Market Paperback edition, 1963.

[Galbraith's book, in the words of a reviewer, helped to codify Western liberalism in the economic realm. Advocating the reduction of stratification of society by wealth, Galbraith argues for policies for social justice using historical data that show that societies that fail in reducing inequity tend to end in revolution or decadence.]

[4] Avinash Bhat: "India's Very Own GPS Is Ready With Seventh Navigation Satellite Launch," *The Hindu*, April 29, 2016. http://www.thehindu.com/sci-tech/science/irnss-launch-indias-own-regional-navigation-satellite-system-takes-shape/article8531388.ece

[5] "Inadequate Judge Strength Is Glaring," *The Hindu*, December 12, 2007. http://www.thehindu.com/todays-paper/tp-opinion/inadequate-judge-strength-is-glaring/article1966009.ece

[6]  "An Overworked Chief Justice TS Thakur Breaks Down in Front of PM Modi," *The Times of India*, April 24, 2016.
http://timesofindia.indiatimes.com/india/An-overworked-Chief-Justice-TS-Thakur-breaks-down-in-front-of-PM-Modi/articleshow/51964732.cms

[7]  Bharat Karnad: *India's Nuclear Policy*, Praegar Security International, Greenwood Publishing Group, CT, USA, 2008.

[8]  John Perkins: *Confessions of an Economic Hit Man*, Penguin Group (USA) Inc., 2004.

# CHAPTER 8

# For the Individual Innovator

The individual innovator is the center forward (or the point guard if you are more used to basketball than soccer) of the game of innovation. This chapter is specifically addressed to him or her, but brief, due primarily to my belief that no one can be taught to become an innovator.

> "It's not clear why anyone should read a book about innovation. There's little evidence people we'd call creative got that way by reading a particular book. Most skills in life are only acquired by work, and to be more creative means to create and learn, rather than merely read." – Scott Berkun [1].

Innovators are internally inspired people with a strong urge to do good, to constantly improve life, and/or to create wealth. That explains one of the reasons why this book's greater focus is on creating an ecology for fostering innovation rather than on creating innovators, or innovations *per se*. The second reason is that the literature on innovation and the various other skills needed for an innovator's success is quite extant.

The experience, narrated in [2], of the great continuous innovator, Intuit Inc., also attests to the fact that learning in the innovation context is indeed by doing. However, I would not go as far as Berkun since I am of the opinion that reading some good books on innovation will, in the least, help one to hone one's skills, to avoid some major pitfalls, and to increase one's chances of success. It also can help one become more adept at identifying opportunities for innovation. To any list of such books, I would also add two that are now classics in the realm of self help books [3], [4]. These deal with personal discipline and the qualities geared to high productivity that are essential for the success of an innovator. More easily than innovating per se, these skills can indeed be learned and developed.

## Research & Innovation

Much confusion exists in the differentiation between research and innovation. The two terms are often used interchangeably. Research and particularly academic type research, to me, is a quest for new knowledge, theories, and inventions for their own sake. The main motivations for it are intellectual curiosity, potential fame, and the attainment of a sliver of immortality in the annals of science and technology. Yet, it behooves researchers to identify opportunities for practical use and commercialization of their research (i.e., to innovate) without letting ideological stands or elitism to get in the way. Being useful in a mundane sense is, after all, as important as being a creative knowledge seeker. Also, the joy of seeing one's efforts being translated into a useful new product or to be put to real use is at least as profound as that yielded by discoveries of an academic type. When it comes to innovation and research, any semblance of elitism of either camp is as absurd as what is described in the following perceptive remark.

> "The happiness of my gardener ... [who] wages a perennial war against rabbits ... is of the same species. ... What joy can we experience in waging war against such puny creatures as rabbits? The argument, to my mind, is a poor one. A rabbit is very much larger than a yellow-fever bacillus, and yet a superior person can find happiness in making war upon the latter." – Bertrand Russell [5].

As noted in an earlier chapter, many of us who are in industrial research in the US are highly cognizant of these facts, and have to be, if only to make it easier for ourselves to find the necessary corporate support and resources to do academic type research.

Unlike research, innovation - as I use the term in this book - is purposive research aimed at solving specific practical problems with economic value. I choose the term 'economic value' instead of 'business value' since much innovation is needed and occurs also in the non-profit arena. The motivations driving an innovator are a desire to be useful in a practical sense, to solve real world problems, to make life better and easier, and/or to create wealth. Thus, when I study a complex probability model like a time varying Brownian motion, including developing novel computational algorithms for it that some (less informed) 'pure' mathematicians may snicker at as 'applied,' I am still doing research of an academic type. However, if I were to create a model based on it to develop good strategies, and software based on that for quantitative trading in stocks or options, then I am being an innovator. That is especially so if the creations are intended for commercial use.

Both research and innovation are equally important, and they should not be viewed as antithetical to each other. Research and innovation (as defined above) can co-exist, be mutually dependent, or be totally disconnected depending on the context and scale. Many of the inventions of companies like Apple and Google are backed by significant (applied and sometime even theoretical) research as were the revolutionary progress in telecommunications and electronics ushered in by AT&T and later by its offshoot Lucent Technologies. Yet, there are many successful innovations, like the pomegranate corer discussed in Chapter 2, that involve little technology and not any significant research of an academic type. The ease of entry offered by such innovations that require not much research, however, is often accompanied by the lack of a moat against cloning, copycatting, or modification and improvement by others. The main point is that the innovator should assess the type of innovation so as to line up needed resources and also create necessary barriers to entry for potential competitors.

## Start with a Problem

A successful innovation does not have to be a complex product that requires some highbrow theory or enormous R&D. It may not involve even a physical product and may just be a new and improved way of doing something. Whatever it is, it is always a good policy to start with a real problem and to understand the actual pain points. An innovation itself becomes valuable only through the value it creates to users. In the commercial arena, one must assess, in addition, whether the pain alleviated is significant enough to induce someone to pay for a solution and if there will be enough buyers to justify one's effort. Certainly, social causes without a profit motive can also be a major catalyst for innovation.

A major mistake of technical people is to start from a novel idea or invention of theirs and then to try to find an application for it. This is like a hammer in search of a nail. For that reason, many technologists make poor business leaders in that they let their fascination for technology cloud their business vision. Except in some rare circumstances where the invention can effect a major shift in the market or the creation of a new market altogether, and has the support of a big organization that can market it effectively, inventions searching for applications rarely go anywhere. Thus, although there are some notable exceptions like 3M's Stick-It notes or Sony's Walkman, it behooves the aspiring individual innovator to start from a problem, whose solution will adequately interest a large number of people to pay a price. At the same time, there is no denying some products like the iPad that have been a huge success even though no user felt any painful need for these products before they were invented and marketed creatively.

Luckily, many problems begging solutions do exist all around us. The sources of opportunities for innovation are so many that Drucker, the author of the classic book on innovation and entrepreneurship [6], has devoted a set of eight chapters covering them. An important lesson I learned from a famous book [7] on stock picking is how our very environment and what we already know can help us in stock picking if we observe critically and with focus. A similar thing is true for identifying innovation opportunities. Thus, it is a worthwhile exercise to understand how many of the gadgets and other inventions we take for granted really came about, because embedded in their history are many interesting and worthwhile lessons for future innovators.

Let me cite two specific examples from my own experience. One relates to the recording, at a customer's request, by a public carrier, of a voice conversation over telephone or other digital media, and various vertical services built on it [8]. I, the primary inventor, was inspired by my being burnt twice. The first time was by an insurance agent who cancelled my collision coverage on all of my cars although I had, over telephone, requested him to do it only on a specific old car. The second related to a long ordeal with a medical insurance company, which claimed that I had not reported the visit of a dependent to an emergency room although it was reported immediately, by phone, by my doctor's office. In the US, an asymmetric situation exists whereby businesses can record every call although individuals have no easy means to make a recording that is admissible in a court of law. My invention was a method by which a telephone company could record a call with attending metadata. It also comprised several vertical services built on top of that. Unfortunately, the ideas never got implemented; more about this will be discussed later.

The second example deals with the start-up company ColorEyeQ, Inc., [9], [10] with which I am involved. It has a set of one-of-a-kind algorithms that can generate, from a set of color measurements, an accurate formula for formulating a color in architectural paints and coatings. It was inspired by the difficulty of matching color accurately by (the ever common) trial and error methods employed by paint stores and manufacturers. In this case, the invention required, in addition to colorimetry and color engineering, several other techniques such as machine learning and some proprietary interpolation and surface fitting tools. We are proud today to have as one of our customers the very inventor of custom color matching for paints, who invented the techniques more than thirty years ago. And, it all started with observing, during routine paint purchases, what goes on in a neighborhood paint store and getting frustrated by the slow convergence of repeated attempts. If you want to be an innovator, start seeing the world around you more critically and with deliberate intent.

Sometimes, opportunities come knocking on the door, but these could be missed easily if one fails either to do some upfront due diligence or to think out of the box, setting aside conventional wisdom. An example of that is a problem of congestion of circuits that even got in the way of calling for emergency services. Work on that by me and several colleagues resulted in a suite of solutions leading to a set of five distinct patents (see [11] for details). The work itself was fulfilling given that the cohort of population affected was estimated to generate roughly 90 calls to emergency services involving a life-threatening emergency each day during the hours when service was adversely impacted.

That meant that even if only a small percentage, say 5%, of the victims could be saved, it would have meant a lot. My involvement came about through the fortuitous invitation by my director to accompany him to a high level meeting with the client to discuss the issues, which the client had (incorrectly) attributed to maintenance activities. Our work also saved much time and money that would have otherwise gone into truck rolls to many homes in the field and a wild goose chase by the labs mining huge amounts of maintenance data. The moral I learned from that engagement is never to leave any stone unturned when solving a real world problem.

### Assess the Market

The assessment of the true addressable market for the innovation and how it may be captured is an exercise that should be undertaken as early as possible. Many innovations witness an early set of enthusiastic takers but never cross the big chasm, called Moore's chasm [12], that exists between those early adopters and the large mass of the address-able market. As we found out with ColorEyeQ, that chasm could indeed be formidable, especially when high-tech solutions attempt to penetrate certain classic industries, which are entrenched in their old ways, or when the "Not Invented Here" syndrome rules high. To state it mildly, our present good fortune with a major paint manufacturer did not come easily, although we had for long the best solution that is also the only one of its class.

The resistance of scientists in accepting new theories, particularly those claiming to make major advances, is well-known [13]. But a similar attitude of skepticism and resistance exists in industry as well. Therefore, an innovator of a product or method that is useful to industry should gain adequate familiarity with the challenges of crossing the infamous Moore's chasm, and not assume that the merit of the solution will bring immediate acceptance. I recommend reading the relevant sections of [14] for some strategic approaches, of which establishing early partnerships with potential opponents is an important one.

The business world can be quite a hostile environment for an innovator who challenges *status quo*. An example is that of the reception given to the electric car in the USA. The ways in which the early attempts at developing them, despite their promise, were squashed by automotive giants are now part of the recorded and verified history of the American automobile industry [15]. Certain innovations like the electric car pose an enormous threat to large businesses using fossil fuel based technologies. Opposition to such innovators can take many forms including regulatory obstacles instigated primarily by vested interests. A recent example is the opposition faced by the carmaker Tesla in many U.S. states when it tried to sell its cars directly to customers [16] bypassing franchised dealers. Some of those rulings are getting reversed in the US [17], thanks to the high sensitivity and awareness of the American consumer to issues related to freedom of choice, and reasonably efficient judicial processes. This is an area where India needs to put in some up-front effort if the attempts at stoking innovation are not to be thwarted by naysayers. The challenges are much bigger in India given its level of corruption, inefficiency of the court system, and the apathy and disempowerment of the consumer. The government should anticipate efforts by the larger players to thwart new innovations and be ahead of the game, and news media should be alert to investigate and publicize wrongdoings. It would also be helpful to help innovators form cooperative organizations that can form a powerful lobby of their own with the ability to share legal and other resources to fight unfair attempts to block innovations.

## Forge Partnerships

The effective implementation of a novel solution, especially in high-tech areas, involves many more steps beyond an idea. Thus, in the example of the congestion controls I mentioned earlier, turning the mathematical solutions into a really usable set required many other steps like simplifying the solution steps so that they could be implemented as controls executing fast within a firmware operating in real time, taking care of various interoperability issues, resolving conflicts with other priorities in routing calls, and the like. Expertise rarely resides in one individual to cover all the technical details, and it therefore becomes necessary to form partnerships and effective teams. These days, even in academic research in many areas, the stereotype of a reclusive scholar working alone has given way to team based collaborative approaches, with teams sometimes spanning diverse subject areas and continents. The really successful innovator is one who has the ability to interface with diverse sets of people and to accommodate differing viewpoints.

## Continuous Learning

An enabler of serial innovation is continuous learning. New technologies are fertile grounds for ever more innovations, and the ones who stay abreast of current developments certainly have an edge over others in being 'ahead of the curve' both in identifying problems and in solving them. In addition to technology related learning, it behooves the innovator to get equipped with a variety of business and management skills in anticipation of a time when the innovation takes off in a big way. Some rudimentary knowledge of intellectual property laws and processes is also helpful in conveying one's ideas effectively to a legal team and in reviewing the output of the latter for technical accuracy.

It is in the interest of industry to encourage continuing education for their employees. The availability of such educational opportunities on the Internet (Coursera, Stanford, MIT, some IITs, for example) makes it much easier today for many aspiring individuals to acquire new skills.

## Protect the IP

It is extremely important to protect the intellectual property underlying one's innovation through appropriate legal means such as copyrights and patents. Large corporations in the US have internal clearance mechanisms to ensure that proprietary intellectual property related information does not get out prematurely in the form of scientific publications or presentations by employees. They are also well set up to help their employees through the maze of filing for patents and copyrights. In India, since R&D is not as well practiced as in the West, many corporations have no policy at all with regard to intellectual property. As India makes a bigger push for more innovations, this gap will have to be filled. For small and medium sized Indian businesses, the situation is even harder in that the legal processes could incur considerable costs. Many issues beg a solution, and finding effective ways of addressing them should occur concurrently as these organizations step up and become drivers of innovation. In the early stages, some governmental help should become available to those innovators determined to be highly promising based on a review by experts. Such help could be in the form of equity investments or loans with highly favorable terms. Some formalized structure, set of rules, and processes need to be evolved as part of the effort to spur innovation in the nation.

## Start-up or Not?

A dilemma that faces an innovator is how to proceed with one's innovative ideas. Should one turn it over to a well-healed existing business or try to do a start-up? This is not an easy question, and the answer would depend on a variety of factors.

When the innovation is by an employee in a corporation and is the result of using knowledge gained directly from one's work or has benefited from the use of corporate resources, it is ethical that the invention be disclosed to the employer. The employee who violates the trust placed in him or her by the employer may not only become unethical, but may even face legal consequences. Laws differ among nations, and in the US even among the states of the Union. Thus, for example, while California gives the employee ample latitude with regard to innovations done using one's own resources and unrelated to one's employment, states like New Jersey are highly favorable to the employers who even write employment contracts stipulating that any and all inventions by the employee will belong to them and fall under the umbrella of 'work for hire.' Many thrust such contracts on temporary employees, contractors, and university collaborators too. The situation in many ways resembles unfair indenture tantamount to intellectual slavery, particularly, when the inventor gets only a trivial fraction of the value generated or none at all. But it is, nevertheless, sanctioned by law in many places, primarily due to crony capitalism.

The employee in a large corporation turning over the invention to the employer, typically, has no control over its fate. It could get implemented in big or small ways, or (like my voice call recording idea that other companies have since made into a successful business) may just gather dust due to a variety of reasons like an unimaginative management, budgetary constraints, the NIH (Not invented here) syndrome, or an over-cautious legal team that is expert and highly creative in imagining the worst possibilities for liability. There is indeed some truth in the assertion,

"One of the causes of unhappiness among intellectuals in the present day is that so many of them ... find no opportunity for the independent exercise of their talents, but have to hire themselves out to rich corporations directed by Philistines, who insist upon their producing what they themselves regard as pernicious nonsense."- Bertrand Russell [5].

The larger the corporation, the larger is the risk of a multi-layered hierarchy, bureaucracy, and rules impeding innovation. That is, unfortunately, the case with many large and old U.S. corporations. Some of the newer corporations like Google, however, seem to have evolved some interesting schemes [18] whereby their engineers are able to enjoy reasonable freedom to innovate and pursue their passions [19] in ways also benefiting the corporation. Such efforts certainly mitigate the type of unhappiness alluded to by Russell leading to better employee retention and also new innovations for the corporation.

The loss of some independence is a price one pays for the security of a paid job essential for some to meet their mundane obligations. It is not to be regretted terribly given some of the comfort and security it offers, particularly during one's formative years, and when it is chosen by free will. The solution perhaps lies more in managing one's own expectations.

## Entrepreneurship

Success as a start-up depends a lot more on one's skills as an entrepreneur than as an innovator. While some innovators like Steve Jobs may also have been great entrepreneurs, it is rare to find people who combine both skills. According to [14] an entrepreneur "is someone who perceives an opportunity and creates an organization to pursue it" and who, with constrained resources, creates value "through innovation, relentless execution, market competition, and sometimes just

[by] doing things better, faster, and cheaper." Note that innovation is just one of the many characteristics marking an entrepreneur. Thus, if an inventor's inclination is to form a start-up, it behooves him or her to either acquire the additional skills and the mindset needed to be a successful entrepreneur, or to partner with one who has a tremendous amount of entrepreneurship.

The multi-lane highways lining successful companies that grew from tiny start-ups may look like offering a joy ride, but their entrance ramps are invariably littered by many wrecked dreams that are out-numbered only by similar wrecks on their bylanes and exit ramps. There is a lot of risk associated with acquiring required funding and resources, turning one's innovation into a product that consumers will like and prefer over others, marketing, and sales and distribution. Invention is but the first step in a long journey turning an idea into a product and a market success. Entrepreneurship and start-ups are a vast topic covering many areas, and the reader may refer to [14] for a good introduction.

## Product Management

Product management also plays a significant role in the success of an innovation in the market. Product management covers many strategic, tactical, and operational issues, all the way from making the product likeable and appealing to various important issues governing branding, product differentiation, positioning and messaging, and coordinating various functions like engineering and marketing to perform together as a symphony. Two great books on product management are those of Cagan [20] and May [21]. These books deal with several aspects of product management in the context of manufactured products and software and are worth the time of an entrepreneur.

A typical start-up cannot afford specialists in many diverse areas, and early founders will have to assume multiple roles simultaneously. Without a serious commitment to long and sustained hard work, and continuous learning far beyond one's original expertise and interest, an innovator cannot succeed as an entrepreneur.

Most importantly, a start-up that remains as a "one trick pony" with just one product is extremely risky. An innovator's goal should therefore be serial innovations that can spread risk across many products or ventures.

## Persistence

Both prolific inventors (like Thomas Alva Edison) and incomparable academic researchers (like C.V. Raman) have attested that their greatest allies were their unflagging persistence, hard work, and determination to succeed. Recall the famous saying, "Genius is one percent inspiration and ninety nine percent perspiration." The world, rarely, offers a free lunch, and if it does, it will hardly be a feast. So, those who wish to feast on success through innovation must, above all, be resolute and willing to go all the way through. It is with that requirement in mind that I cited the two self-help books [3], [4] very early in this chapter.

Hard as it may seem, let us be enthused by the fact that many have successfully traversed the path of innovation, and that many of the giant oaks of today were indeed born as tiny acorns. My constant reference on start-ups [14] ends with a quote from Victor Frankl,

> "Don't aim at success – the more you aim at it, and make it a target, the more you are going to miss it. For success, like happiness, cannot be pursued; it must ensue."

Our own Indian equivalent from Bhagavad Gita "*karmanyēvā-dhikārastē mā phalēshu kadāchana...* (You only have a right to action and not to the fruits thereof)", BG 2-47, is even stronger, and may it, together with the exhortation "*uddharēt ātmanātmānam* (elevate the self by the self)", BG 6-5, form the motto of the large army of innovators like you that we need to propel India into the developed world!

## Notes and References

[1]  Scott Berkun: "The 5 Best Books on Innovation EVER," Feb 7, 2012.^http://scottberkun.com/2012/the-5-best-books-on-innovation-ever/^

[2]  Roger L. Martin: "The Innovation Catalysts," in *HBR's 10 Must Reads on Innovation*, Harvard Business Review Press, 2013.

[3]  Napoleon Hill: *Think and Grow Rich*, Revised & Expanded by Dr. Arthur R. Pell, Jeremy P. Tarcher/Penguin, ISBN 1-58542-433-1, 2005.

[4]  Stephen R. Covey: "*The 7 Habits of Highly Effective People: Powerful Lessons in Personal Change*," Free Press, Simon & Schuster, Inc., 2004.

[5]  Bertrand Russell: *The Conquest of Happiness*, First Published in Great Britain by George Allen & Unwin, 1930.

[6]  Peter Drucker: *Innovation and Entrepreneurship*, Harper Collins, 1993.

[7]  Peter Lynch: *One Up on Wall Street*, Simon & Schuster, NY, ISBN 0-7432-0040-3, 2000.

[8]  Byers, et al: "Method for providing a phone conversation recording service," U.S. Patent 6,987,841, Jan 16, 2006 in continuation of the application "Authentication Of Communications with Recordings Storage and Playback", V. Ramaswami, Oct 31, 2001.
http://patft.uspto.gov/netacgi/nph-Parser?Sect1=PTO2&Sect2=HITOFF&p=1&u=%2Fnetahtml%2FPTO%2Fsearch-bool.html&r=10&f=G&l=50&co1=AND&d=PTXT&s1=Ramaswami.INNM.&s2=Vaidyanathan.INNM.&OS=IN/Ramaswami+AND+IN/Vaidyanathan&RS=IN/Ramaswami+AND+IN/Vaidyanathan

[9]  http://www.coloreyeq.com/

[10] Kicha Ganapathy, A talk presented at the IIM, Bangalore Alumni meeting on Disruptive Technologies, May 2014. https://www.youtube.com/watch?v=BHF1qi2OTew

[11] Ramaswami, V et al: "Assuring emergency services access: providing dial tone in the presence of long holding time internet dial-up calls," 2004 INFORMS Wagner Prize Finalist Paper, *Interfaces*, 35, (2005), pp. 411-422.

https://www.researchgate.net/profile/V_Ramaswami/
publications?sorting=newest&page=2

[12] Goeffrey Moore: *Crossing the Chasm: Marketing and Selling High-Tech Products to Mainstream Customers*, Harper Collins, New York, 2002.

[13] Thomas S. Kuhn: *The Structure of Scientific Revolutions*, The University of Chicago Press, 50th anniversary edition, 2012.

[14] Gregg Fairbrothers & Tessa Winter: *From Idea to Success, The Dartmouth Enrepreneurial Netowork's Gude for Start-ups*, McGraw-Hill, 2011.

[15] "Who Killed the Electric Car?" Sony Pictures Classics, http://documentaryheaven.com/who-killed-the-electric-car/

[16] "New Jersey to Tesla: You're Outta Here," *Forbes*, http://www.forbes.com/sites/michelinemaynard/2014/03/11/new-jersey-to-tesla-youre-outta-here/ - 71a4388e5b0b

[17] "New Jersey Finally Lifts Ban on Tesla Sales," *Huffington Post*, http://www.huffingtonpost.com/2015/03/18/new-jersey-tesla-ban-lifted_n_6896896.html

[18] Kathy Gersch: "Google's Best New Innovation: Rules Around '20% Time'," Forbes, Aug 21, 2013. http://www.forbes.com/sites/johnkotter/2013/08/21/googles-best-new-innovation-rules-around-20-time/ - 583482a68b85

[19] David Goldman: "Google Gives 20% to Japan Crisis," CNN Money, Mar 17, 2011, http://money.cnn.com/2011/03/17/technology/google_person_finder_japan/

[20] Marty Cagan: Inspired: *How to Create Products Customers Love*, *SVPG Press*, an imprint of Silicon Valley Product Group, Inc., CA, ISBN:978-0-9816904-0-8, 2013.

[21] Chris Vander Mey: *Shipping Greatness: Practical Lessons on Building and Launching Outstanding Software, Learned on the Job at Google and Amazon*, O'Reilly Media, Inc., ISBN: 978-1-449-33657-8, 2012.

# 'Breathes there the man'

## Sir Walter Scott

Breathes there the man, with soul so dead,

Who never to himself hath said,

This is my own, my native land!

Whose heart hath ne'er within him burn'd,

As home his footsteps he hath turn'd,

From wandering on a foreign strand!

If such there breathe, go, mark him well;

For him no Minstrel raptures swell;

High though his titles, proud his name,

Boundless his wealth as wish can claim;

Despite those titles, power, and pelf,

The wretch, concentred all in self,

Living, shall forfeit fair renown,

And, doubly dying, shall go down

To the vile dust, from whence he sprung,

Unwept, unhonour'd, and unsung.

# Epilogue

I am a Non Resident Indian. I know that the NRI is held in disdain in many circles in India and given epithets worse than the *mlecha* (outcast) that A.L. Basham, the author of the wonderful book, *The Wonder That Was India*, (whose only fault, I think, is the use of the past tense in its title) felt obligated to call himself. But I am also an NRI with a deep connection to India and some great debts to repay.

My grandfather Dr. T.S. Anantha Padmanabhan of Badagara, Kerala resigned a residency because he felt outraged at being asked, "Who the bloody devil do you think you are?" when he politely pointed out that his British boss at Stanley Hospital, Madras had made a wrong diagnosis. In his late twenties and inspired by Mahatma Gandhi, he exchanged his stylish British attire of shorts, hat, and knee-high stockings permanently for hand-spun white cotton. He marched to sea repeatedly to make salt to defy British salt tax laws and faced many official harassments. He spent most of his enormous wealth earned as a private practitioner in support of India's freedom movement and died a pauper. And he taught me the wonderful lines of "Breathes there the man .... " and instilled in me an uncompromised love for my motherland. His father, Tarakad Subrahmanya Iyer, resigned his post as Collector in the British Raj in acquiescence to his son's passions just as Motilal acquiesced to Jawaharlal, and spent his remaining days in the voluntary service of the Theosophical Society, Madras. A paternal uncle, T.R. Viswanathan, ran off to fight the British in the army of Subhash Chandra Bose. A maternal uncle, T.A. Parameswaran, spent his youth disrupting British troop trains and cutting official telegraph cables, and later in independent India became a decorated Police Superintendent. These are my unsung heroes, and it is in their honor that I offer this book.

It is sad that even someone like Dr. Abdul Kalam should relegate scientists like me as those "who leave this country at their first opportunity to earn more money abroad." How could he be unaware of the acute economic reasons exacerbated by the absence of adequate opportunities and (reverse) discrimination based on caste that still prevail and continue to drive many away?

Be that as it may, the NRI contribution to India is very significant. Their exemplary contributions as scientists and professionals, and now as innovators, and makers and movers of global businesses, have elevated significantly the perception of the Indian from the stereotype of a "coolie" and a "snake charmer." Many of those NRIs are my friends, and I honor them along with my and their children who, as U.S. citizens, will always have a soft corner for India and will proudly uphold their Indian culture and heritage.

I thank the J.N. Tata Educational Foundation and the K.C. Mahindra Foundation whose interest-free loan scholarships paid for my clothes and airfare to the USA, but for which I would have had to forego a Ph.D. Fellowship from Purdue that came with no strings attached. I thank my wonderful teachers of whom I must mention two by name: Professor K. Balasubramanian of Loyola College, Madras and Dr. K.N. Venkataraman of the University of Madras who went many an extra mile to encourage me. I thank Rev. Fr. Francis, Principal of Loyola College, who wrote many 'slips' to let me pay my term fees whenever I could and never made me feel embarrassed for asking. Finally, I thank my mother and my grandmother who happily sold even the last pieces of their meager jewelry for my higher education.

This book reflects not just my experiences and readings, but even more my passion for India. In case some passages have offended you, it is not because, to paraphrase unabashedly a famous line, "that I love you less, but that I love India more." Do get angry, and I indeed want you to, but please direct your anger in ways that would lead this book to becoming irrelevant and eventually obsolescent. Thank you for staying with it all the way up to this page.

**V. Ramaswami**

# BIBLIOGRAPHY

Bijapurkar, Rama: *We Are Like That Only – Understanding the Logic of Consumer India*, Penguin Books, India, 2007.

Brandt, Ronald S (Editor): *Education in a New Era*, Association for Supervision and Curriculum Development, Alexandria, VA, USA, 2000. ISBN: 0-87120-363-4.

Cagan, Marty: *Inspired: How to Create Products Customers Love*, SVPG Press, an imprint of Silicon Valley Product Group, Inc., CA, ISBN:978-0-9816904-0-8, 2013.

Carlson, Lucas: *Finding Success in Failure: True Confessions from 10 Years of Startup Mistakes,* Craftsman Founder, 2015, ISBN 978-0-9960452-2-3.

Cashman, Sean Dennis: *America in the Gilded Age*, New York University Press, 1984.

Chengappa, Raj: *Weapons of Peace,* Harpercollins Publications India Pvt. Ltd., 2000.

Cohen, Stephen P: *India: Emerging Power,* Brookings Institution Press, Washington, DC, 2001.

Covey, Stephen R: *The 7 Habits of Highly Effective People: Powerful Lessons in Personal Change,"* Free Press, Simon & Schuster, Inc., 2004.

Drèze, Jean & Sen, Amartya: *An Uncertain Glory, India and Its Contradictions,* Princeton University Press, Princeton & Oxford, 2013.

Drucker, Peter: *Innovation and Entrepreneurship,* Harper Collins, 1993.

Fairbrothers, Gregg & Winter, Tessa, : *From Idea to Success, The Dartmouth Enrepreneurial Netowork's Gude for Start-ups,* McGraw-Hill, 2011.

Feynman, Richard A, *'Surely You're joking, Mr. Feynman!' Adventures of a Curious Character, 1985*, Bantam Books, New York.

Finkel, Robert: *The Masters of Private Equity and Venture Capital,* McGraw Hill, 2010.

Galbraith, J.K: *The Affluent Society,* ISBN 0-395-92500-2, Mariner Books, The Mass Market Paperback Edition, 1963.

Gertner, John: *The Idea Factory: Bell Labs and the Great Age of American Innovation,* Penguin Press, 2012.

Grubman, James: *Strangers in Paradise, How Families Adapt to Wealth Across Generations,* ISBN-13: 978-06-15894355, Paperback, Family Wealth Consulting, www.jamesgrubman.com, 2013.

Guha, Ramachandra: *India After Gandhi: The History of the World's Largest Democracy,* Harper Collins Publishers, 2007.

*HBR's 10 Must Reads On Innovation,* Harvard Business Review Press, Boston, MA, 2013.

Hill, Napoleon: *Think and Grow Rich,* Revised & Expanded by Dr. Arthur R. Pell, Jeremy P. Tarcher/Penguin, ISBN 1-58542-433-1, 2005.

Hunter, W.C., Kaufman, G.G. & Krueger, T.H: *The Asian Financial Crisis: Origins, Implications and Solutions,* Kluwer Academic Publishers, 1999.

Kalam, A.P.J. & Rajan, Y.S: *India 2020, A Vision for the New Millennium,* Penguin Books India, New Delhi, 1998.

Kalam, A.P.J. (with Arun Tiwari): *Wings of Fire, An Autobiography,* Universities Press, 1999.

Karnad, Bharat: *India's Nuclear Policy,* Praegar Security International, Greenwood Publishing Group, CT, USA, 2008.

Kennedy, Donald: *Academic Duty,* Harvard University Press, 1977.

Kiyosaki, Robert T: *Rich Dad Poor Dad: What the Rich Teach Their Kids About Money That the Poor and Middle Class Do Not,* Warner Books, NY, 2011.

Kuhn, Thomas S: *The Structure of Scientific Revolutions,* The University of Chicago Press, 50th anniversary edition, 2012.

Kumar, Shail: *Building Golden India - How to Unleash India's Vast Potential and Transform Its Higher Education System. Now,* ONS Group Press, Fremont, 2015. ISBN: 978-0-9966168-0-5.

Lusk, John & Harrison, Kyle: *The MouseDriver Chronicles: The True-Life Adventures of Two First Time Entrepreneurs,* Basic Books, 2002, ISBN-10: 0-7382-0801-9.

Lynch, Peter: *One Up on Wall Street,* Simon & Schuster, NY, ISBN 0-7432-0040-3, 2000.

Malhotra, R & Neelakandan, A: *Breaking India: Western Interventions in Dravidian and Dalit Faultlines,* Infinity Foundation, Princeton, NJ. ISBN: 978-1-937037-00-0, 2011.

Mey, Chris Vander: *Shipping Greatness: Practical Lessons on Building and Launching Outstanding Software, Learned on the Job at Google and Amazon,* O'Reilly Media, Inc. ISBN: 978-1-449-33657-8, 2012.

Moore, Goeffrey: *Crossing the Chasm: Marketing and Selling High-Tech Products to Mainstream Customers,* Harper Collins, New York, 2002.

Mukerjee, Madhusree: *Churchill's Secret War: The British Empire and the Ravaging of India During World War II,* Basic Books, A member of the Perseus Book Group, USA, Paperback, 2010.

Panagaria, Arvind: *India: The Emerging Giant,* Oxford University Press, 2008.

Perkins, John: *Confessions of An Economic Hit Man,* Penguin Group (USA) Inc., 2004.

Peter, Lawrence & Hull, Raymond: *The Peter Principle: Why Things Always Go Wrong,* Harper Collins, 2009.

Prahalad, C.K: *Fortune at the Bottom of the Pyramid: Eradicating Poverty Through Profits,* Pearson Education Inc., Prentice Hall, 2005.

Romans, Andrew: *The Entrepreneurial Bible to Venture Capital,* McGraw-Hill, 2013.

Russell, Bertrand: *The Conquest of Happiness,* First Published in Great Britain by George Allen & Unwin, 1930.

Sengupta, Hindol: *Recasting India, How Entrepreneurship is Revolutionizing the World's Largest Democracy,* Palgrave Macmillan Trade, New York, 2014.

Stanley, T.J & Danko, W.D: *The Millionnaire Next Door: The Secrets of America's Rich,* Pocket Books, NY, 1996.

Stiglitz, Joseph: *Globalization and its Discontents,* W.W. Norton & Company, NY, 2002.

Stronge, James H: *Qualities of Effective Teachers,* Association for Supervision and Curriculum Development, Alexandria, VA, USA, 2002. ISBN: 0-87120-663-3.

Subramanian, Arvind: *India's Turn: Understanding the Economic Transformation,* Oxford University Press, 2008.

Tellis, Ashley J: *India's Emerging Nuclear Posture: Between Recessed Deterrent and Ready Arsenal,* RAND Corporation, ISBN:0-8330-2781-6 (pbk), 2001.

Temin, Peter & Galambos, Louis: *The Fall of the Bell System,* Cambridge University Press, 1987.

Trevelyan, George Otto, *The Life and Letters of Lord Macaulay,* "Longmans, Green & Co, London, 1876.

Vander, Chris May: *Shipping Greatness: Practical Lessons on Building and Launching Outstanding Software, Learned on the Job at Google and Amazon,* O'reilly Media Inc., ISBN: 978-1-449-33657-8, 2012.

Vise, David A: *The Google Story,* Bantam Dell, Random House Inc., New York, 2005.

Warrier, A.G.K: *Srimad Bhagavad Gita Bhasya of Sri Sankaracarya,* Sri Ramakrishna Math, India. ISBN 978-81-7823-507-3.

# INDEX OF PERSONS

# GENERAL INDEX

www.ingramcontent.com/pod-product-compliance
Lightning Source LLC
Chambersburg PA
CBHW060007210326
41520CB00009B/843